RENOVATE

Metropolitan Home

RENOVATE

WHAT THE PROS KNOW ABOUT GIVING NEW LIFE TO YOUR HOUSE, LOFT, CONDO OR APARTMENT

FRED A. BERNSTEIN

CONTENTS

INTRODUCTION

For more than 20 years, the favorite sport of *Metropolitan Home* magazine readers has been renovation. Some of our longtime subscribers have renovated more than one house since they signed on with us; others have reinvented the same place more than once. The editors dedicate at least one issue of the magazine a year to it, and many stories in all of the other issues. The "Renovation Issue" is always a newsstand bestseller that generates extraordinarily passionate reader response.

Personally, that enthusiasm comes as no surprise. I've renovated two 19th-century city townhouses and one 1932 suburban treasure, and I am currently tackling a late-18th-century farmhouse, a project that has already gone on for several years. Some of my efforts have been minor—restoration of glorious parquet floors, for example—and less than glamorous, like replacing lead plumbing. But some of my projects have been major, involving the removal or shifting of walls. At one point I added a kitchen that, as one friend put it, made the rest of the house obsolete "You sleep in the bedroom," she notes, "and spend all the rest of your time in the kitchen." That kitchen is 18 years old now, and I still marvel at it every time I walk into the room. What more could I ask for?

Obviously the rest of America is busy perfecting its home dreams too: There are now more houses being renovated than built from scratch—and this has happened during a housing market boom. But renovation makes perfect sense for a lot of reasons:

* A surfeit of soul. There's always some elusive mystery in their past: Why did someone put that closet there? How huge were the trees that produced the wide board floors? Who cooked the meals in that cavernous fireplace?

∗ Location, location, location. Old houses—even 1970s ranches, which now qualify as elderly—are closer to city centers and often in beautifully established neighborhoods.

∗ The recycle ethic. It's obviously far more environmentally friendly to reclaim existing housing than to build new, especially when the new has supplanted a farm or a wildlife corridor.

∗ Character mix. Combining old and new creates compelling visual drama: the play of old pine paneling against sleek glass walls, sheer new cabinets set off by antique mullioned windows, the wondrous surprise of a traditional exterior and an open, airy modern interior. The possibilities are endless.

∗ The learning curve. Fred Bernstein, the author of this book and a long-time *Met Home* contributing editor, says that the learning curve of his first renovation was so steep, he can't wait to start a new renovation, just to apply what he learned. I think that each renovation teaches many, many lessons. But there's enormous satisfaction in taking what you've learned and building on it the next time around.

Because renovation is as much about problem-solving and process as it is about design and choices, both the challenge and the reward are huge. Hopefully this book will help with ideas, advice and inspiration—wherever you live now and however you'd like to live in the future.

Donna Warner

Donna Warner
Editor in Chief, *Metropolitan Home*

CITY HOUSES

FLOOR
SHOW

WHAT THEY HAD A HOUSE THAT WAS DARK AND SO CHEAPLY CONSTRUCTED THAT IT SHOOK IN THE WIND—BUT OFFERED FABULOUS PACIFIC OCEAN VIEWS.

WHAT THEY WANTED A PLACE TO KICK BACK AND CHILL, WHERE FRIENDS WOULD COME TO DINNER AND NEVER WANT TO LEAVE, THANKS TO A PADDED ROOM FOR LISTENING AND LOUNGING.

HOW THEY GOT IT The location—at the top of a hill in the center of San Francisco—was staggering, but the building was anything but. "It was a dump," says Shane Ginsberg of the 1970s hill-hugger. "It literally shook in the wind." Not even the Pacific Ocean got a break. On the main level, containing an enclosed kitchen and an L-shaped living/dining room, the ceiling sloped down to sliding glass doors, effectively letterboxing the views.

With the help of Nilus de Matran, a San Francisco architect, Shane and his wife, Kate, removed unnecessary walls and installed a new steel beam that would support a level, 15-foot-high ceiling. "Standing there, with the views and the light, the wind and the fog," says Kate, "it all made sense. There was nothing between us and the ocean."

Lovers of modern architecture, the couple asked de Matran to avoid fussy details. He says, "I don't hide things away, like a true minimalist, but I work to simplify what's there." The stairway is made of metal grates, rather than solid metal or wood, so that light and air would filter through. Exterior walls (and even some of the interior surfaces) are made of Hardieboard, a kind of concrete panel that suggests raw stone (see box, page 14). That gives the house an industrial quality, which Kate and Shane then balanced with feminine shapes—hourglass pendant lights, an arched Achille

Opposite: Concrete panels give the house an industrial look—which curvaceous light fixtures (by Karim Rashid) and dining chairs (by Jasper Morrison) soften. The kitchen island is Carrara marble over dark-stained walnut; it stands where a wall had needlessly enclosed the kitchen.

before

What the Pros Know About
Concrete Siding

For the outside of the house, plus several key interior walls, Nilus de Matran chose concrete panels, which are the same size (four-by-eight feet) and about the same weight as plywood sheets. The boards can be saw-cut and screwed into place. **But unlike plywood, they may last 100 years without any maintenance.** The boards' concrete surface can be stained or waxed or left untreated (which was Shane and Kate's choice—they like the material's resemblance to rough stone). Since concrete expands in hot weather, de Matran says, it's important to drill the holes a little bigger than the screws used to secure the panels, and to make sure there are gaps between the sheets. He filled the joints with silicone for waterproofing.

What the Pros Know About
Terrazzo Floors

Terrazzo consists of bits of stone suspended in an adhesive; the floor is poured, then polished in place. Shane Ginsberg associates terrazzo with the late 1950s and early 1960s, when the line between indoors and outdoors was often blurred. In fact, the material has been improved since its mid-century heyday; acrylic has replaced concrete as the adhesive. **That means the floor is less likely to crack as the house settles.** And it means the floor can be poured thinner, and hence lighter—so you don't need extra structure to support it. A terrazzo floor should last almost forever, and, with flecks of gray and black marble (as in the Ginsbergs' house), it hardly shows dirt. The only downside: At $25 or more per square foot, it's pricier than many other floors.

Opposite: The front of the house (far left) reveals little of its true size. Had they built higher, the Ginsbergs would have blocked their neighbors' views. The metal gate leads to a small entry courtyard. This page: The dining table is by architect Nilus de Matran; the painting, by his wife, Jennifer Morla.

Castiglioni floor and Jasper Morrison chairs with rounded seats and backs.

The main level features a fireplace of cold-rolled steel (surrounded by cabinets of dark-stained walnut), and a kitchen island so big that, Kate says, "I could perform surgery on it." The island floats in a sea of white terrazzo—900 square feet of it (see box, page 14).

De Matran created computer simulations of the rooms, with beds, sofas and tables—and then the couple shopped for pieces that best matched the simulations. But there's one room that needed almost no furniture: a soundproof white vinyl lair with a built-in upholstered platform for lounging and listening to music. (De Matran had an upholsterer make vinyl panels, which he then simply glued to the walls.) Under the platform, a trundle bed with a queen-size mattress pulls out to sleep guests. Unlike so

many architects, who do modern buildings only to see their clients install frumpy furniture, de Matran admires the Ginsbergs' discipline. "Kate and Shane," he says, "were willing to go all the way."

RENOVATION DREAM As newlyweds rebuilding a house from top to bottom, the Ginsbergs had reason to be apprehensive. "Renovations can end relationships," says Kate. It was Shane's idea to give himself, his wife, their architect, Nilus de Matran, and their contractor, John Hakewill, three "silver bullets" each—meaning that each had three absolute vetoes. "The idea," says Kate, "was to force you to decide what you just didn't like a little, and what you really couldn't live with. Something you really couldn't live with, you could overrule." Perhaps because they had the bullets, not a single one was fired.

Above: In the office, on the lowest level of the house, Karim Rashid's stacking chairs surround a table by Nilus de Matran. Opposite: A padded room (for listening and lounging) nestles under the garage.

Garage   Entry   Living Room/Dining Room/Kitchen   Top
                                                       1st Level

Padded
Room

Bath   Master   Middle
                              Bedroom   2nd Level

Mezzanine   Office   Lower
                                                 3rd Level

The couple's ethereal bedroom "makes us feel like we're sleeping in a cloud," Kate says. Translucent closet doors were made of ¼-inch-thick sandblasted glass, mounted in aluminum channels by cabinetmaker George Slack. When Shane is away, Kate hosts "glamour parties" for her girlfriends in the spa-like bathroom.

UNCOMMON BRAHMIN

WHAT THEY HAD A SHELL OF AN 1840S HOUSE ON BOSTON'S BEACON HILL, WITH 70 WINDOWS REQUIRING RESTORATION, AN INTERIOR WITH NOTHING BUT ROUGH FLOORING AND A DEVELOPER IN A BIG HURRY.

WHAT THEY WANTED A FAMILY HOME WHERE FLOWING SPACES (NOT TINY ROOMS) FILL THE FOUR LEVELS AND WHERE MATERIALS ARE AS WARM AS THE RADIANT HEAT RISING FROM THE MAHOGANY FLOORBOARDS.

Opposite: The house, on
Boston's historic Louisburg
Square, faces a corner—
giving it more than its fair
share of windows. It
had been chopped up into
apartments, and "we were
returning it to its original
use as a single-family
home," says David Hacin.
Above: Hacin designed
the ornate stairway and
the master bedroom's
Shaker-simple built-ins.

HOW THEY GOT IT The best way to renovate a house, Meg Gordon says, "is to live in it for a while before you begin making changes. Then you'll know what rooms you're in the most. And you won't spend a lot of money on furniture and fixtures that you'll hardly ever use."

Good advice, but Gordon wasn't able to follow it. When she and her husband, James, bought their house on Boston's Beacon Hill, there was nothing to the building but its 1840s brick shell. The seller was willing to renovate it to the couple's specifications, but on a schedule: The Gordons had just two months to sit down with their architect—the versatile David Hacin—and pick out everything they needed, from appliances to light fixtures to door knobs.

But the fact that the house had been gutted gave them freedom. "In most houses on Beacon Hill the rooms are small, and they're cluttered with too much furniture," says Hacin. But in this case, "the clients are a new breed of Beacon Hill owners—they didn't want to get caught up in re-creating a historic style."

Instead, Hacin came up with a layout composed of large, open spaces (including a kitchen/family/dining room that fills an entire level). And he was able to choose materials, including a mahogany flooring, that exude warmth—as much as the coils below the floor radiate heat (see box, page 26). "I'm a big fan of bringing in colors through materials," says Hacin, who figures that when the architecture is colorful, the furniture can be more subdued. "Besides," he says, "the mahogany—which wouldn't have been used originally—gave the whole house a twist."

But this being Beacon Hill, there were limits to how adventurous the couple could be. The house's 70 double-hung windows had to be restored to precise specifications, following the neighborhood's historic preservation guidelines. "Just getting the color of our front door approved took four meetings," says Meg (it's black). Luckily, James

Gordon—the developer of a planned windmill farm in Nantucket Sound—
has experience with complicated projects.

The windows suggested the house's basic demeanor—colonial—
but that didn't mean everything had to be as it would have been in Paul
Revere's day. Hacin developed details that were colonial in spirit but spare
enough that the contemporary furniture chosen by designer Manuel De
Santaren (including pieces by Christian Liaigre) wouldn't feel out of place.

The Gordons' most significant decision was to put the kitchen on the
third floor of the house. Most townhouse kitchens, Meg observes, are at
ground level, where light and views are scarce. From the third floor, she
could walk out onto a roof deck with spectacular views of the Boston
Public Garden and Back Bay.

But having a kitchen on the third floor meant climbing lots of stairs.
Hacin convinced the couple to install a tiny elevator—which, Meg says,
seemed extravagant but now seems essential. "I thought I would be

shopping for two people my whole life." But with two children, Meg is riding the elevator with groceries for four.

RENOVATION DREAM By locating their kitchen on the third floor (with bedrooms above and living room below), the Gordons ensured that it would always be at the heart of the house. James, a lover of technology, asked for state-of-the-art equipment. Meg, a chef, wanted professional appliances but was uncomfortable with a kitchen too modern for the historic building. The compromise was a kitchen from Germany's SieMatic, which architect Hacin likens to a Porsche engine in a classic car body. Cabinets are hand-painted wood, and countertops are teak. (Hacin says teak requires maintenance—including regular oiling—and will show marks, "but part of the appeal of teak is that it does give you a weathered, distressed feeling over time.") Hacin's layout included a walk-in pantry, allowing him to limit the number of upper cabinets (a move that will make any kitchen feel spacious). The shape of the room—long and narrow—provides the intimate work spaces that Meg enjoys. And the linear layout creates a natural division between cooking and serving areas.

Opposite: The mantel is cast stone, from a Texas company called Old World Stoneworks. Hacin says, "It has traditional proportions but a streamlined feel."

before

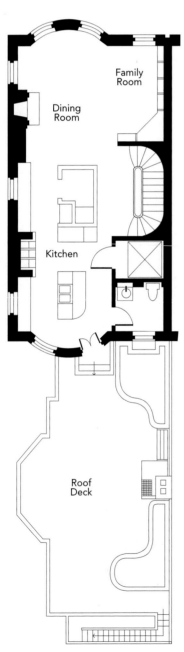

What the Pros Know About
Radiant Heat

Pioneering architects—including Frank Lloyd Wright—were partial to radiant heat, in which coils filled with hot water run under the floor. The coils heat the surface, which in turn warms the room. The system gives architects the freedom to create walls of glass, uninterrupted by vents or baseboards. But even in a house with conventional windows and plenty of room for baseboards, radiant heat "is a dream," says Meg Gordon. According to Meg, "It takes a while to get the settings right. The company had to tweak it, but once it's set, you can forget it. You don't have to deal with forced air drying you out all winter. And it's wonderful to walk around in bare feet in New England in the dead of winter." Radiant systems tend to be expensive, and there's always the danger of leaks. Says Meg, "Of course, you're told you can't nail anything into the floor." **Some designers say that, because of the possibility of leaks, radiant is best for single-story houses (where only the basement stands to suffer).** But Meg says the one small leak she experienced was "no big deal" and that she "would never want to live in a house without radiant heat again." James Gordon, a developer of alternative energy sources, says that radiant requires about at least 30 percent less energy than other systems, because the water only has to be heated to 105 degrees—not the 180 degrees of conventional hot-water heat.

Throughout the house, architect Hacin developed streamlined versions of authentic period details. "The proportions of the crown moldings and paneling are accurate, but I stripped them down to basics," he says.

WORKING

CLASSIC

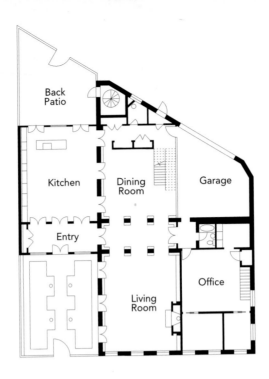

Back
Patio

Kitchen

Dining
Room

Garage

Entry

Living
Room

Office

WHAT THEY HAD A SECTION OF AN OLD CHICKEN-PROCESSING PLANT, WITH BRICK WALLS, A SMOKESTACK AND A ROOF THAT COULD SUPPORT A 747.

WHAT THEY WANTED A HOME THAT MANAGES TO BE AS GRAND AS IT WAS GRITTY, AND AN OFFICE JUST STEPS FROM THEIR BEDROOM, FOR "THE WORLD'S SHORTEST COMMUTE."

Previous spread: In the 24-by-28 foot living room, architect Larry Booth covered a frame fireplace in sheets of steel, in homage to the house's industrial past. Opposite: Double doors lead from the dining room (in the original building) to a newly constructed kitchen.

HOW THEY GOT IT When Paul Zucker, a Chicago real estate developer, was assembling a site in the Chicago neighborhood of Lakeview for a planned mixed-use development, he came upon a brick building that had housed the boiler of an old chicken-processing plant. His wife, Susan, thought the couple ought to turn the structure into their new house. But the condition of the building nearly scared her off: Outside, it was graffiti-covered; inside, it was dark, drippy and smelly. Pigeons, she recalls, had taken over.

Then, too, the building didn't exactly suggest a cozy domestic layout. "This wasn't the kind of renovation where you say, 'Here's the kitchen; let's put in new cabinets. There's the bathroom; let's buy a new sink,'" Susan says. Adding to the challenge: The building would contain not just the couple's home, but offices for their real estate company. Paul's goal: "the world's shortest commute."

Two things clinched the deal. The building's concrete-beamed ceiling was so strong, Susan recalls, "that someone said you could park a 747 up there. So I figured, if you can have a 747, you can have a pool. Paul has a shoulder injury, and keeping the shoulder moving, by swimming, is the best therapy." Then Larry Booth (of Booth Hansen Associates), among Chicago's most respected architects, came to see the building. On the spot, he told the Zuckers that he felt inspired by the building.

The couple gave Booth constraints. They wanted to add a second floor, but insisted the stairway be where there was already a cutout for a

before

Opposite: The oversize
grid of the concrete floors
reflects the building's
industrial scale. When
Booth warned Susan
Zucker that the floor, with
fewer control joints than
normal, would likely crack,
she replied that she
liked the idea of the
building showing its age,
just like a person.

What the Pros Know About
Medium-Density Fiberboard

Medium-Density Fiberboard—MDF—has long been
the stuff of inexpensive kitchen cabinets, but architects
are beginning to cast the material (composed of
pressed wood chips and resin) in leading roles. Here, architect Larry Booth wanted to give the windows and
columns of the Zucker house dramatic outlines. Some
of the elements of his design are 9 feet wide and 12
feet high. "We never could have found lumber that
big," Paul Zucker says. More to the point, Booth's au-
dacious detailing requires precision. With MDF, **"you
don't get warping or twisting," says Paul. You also
get dimensional stability, meaning, unlike wood,
MDF doesn't shrink.** When you go to the higher
grades of MDF, he adds, "it's a very smooth surface—
smoother than drywall would be." Paul, a real estate
developer, is continuing to work with the material. "In a
lot of our high-end residential developments, we're using MDF for baseboards and other moldings, because
you get a more consistent finish than with wood."

skylight in the roof of the existing building. "Was it the most logical place for a stairway?" Susan asks. "Maybe not, but you work with what you have." (A second stairway was inserted in the old brick smokestack.) The building's odd shape created "a lot of dissonance," Booth says. His goal was to substitute "harmony and rhythm and proportion." To do that, Booth cut large openings in the brick walls, inserting windows in a regular pattern that suggests a classical sense of order. To emphasize the new windows, Booth framed them in "boxes" of medium-density fiberboard (MDF). That gives the windows the proportions of French doors, even when—for privacy reasons—the glass is above shoulder level.

The remaining brick walls were covered in Sheetrock, but with a Boothian touch: The Sheetrock stops, leaving the top third of the walls exposed. The goal, Booth says, "is to emphasize solidity. You've got a heavy ceiling—it can't look like there's just drywall holding it up." Overhead, square skylights follow the grid created by Booth's windows. As a result of the rigorous geometry, "people sense a wholeness when they walk in."

The Sheetrock doesn't rest on the brick, but on studs that hold it several inches away. In the gap, Booth installed fluorescent fixtures; light seems to come from out of nowhere. Booth's idea, says Susan, "is that the house would light itself." And it does, like a beacon.

The kitchen (above and opposite), with seamless granite countertops, is contained in a wing designed by Booth to complement the original building. New exterior walls (middle) were clad in zinc, which comes in rolls and was installed with overlaps, like shiplap siding.

before

What the Pros Know About
Indoor Pools

Having a pool inside is something many people dream of; having a pool upstairs is something few would even consider. Then again, not many people have the kind of ceiling beams (see "before" photo) that would allow them to hoist a 40-foot-long fiberglass pool ("a giant bathtub," Paul calls it) onto the roof, build a new master bedroom suite around it and, finally, fill it with nearly 8,000 gallons of water. **Paul (whose office is directly below the pool) says there's never been a leak.** One problem with indoor pools, he says, is that the humidity and odors end up spreading throughout the house. The way to avoid that is to do what Paul did, and give the pool area a separate HVAC system. **The system in the pool area extracts moisture from the air and returns heat gained in that process to the pool.** As a result, "we have never needed a separate pool heater," says Paul. Still, the system "raises your electric bill significantly." And though the swimming pool provides stunning vistas through the glass wall of the couple's bathroom, he says that it's usually covered—because when it's uncovered, "there's so much moisture in the air that running the system gets really expensive." Instead of chlorine, Paul chose a hydrogen peroxide purification system, which "produces less odor and is easier on your hair and your skin and your clothes than chlorine." But (there had to be a catch) it costs about a third more than a chlorine system.

VISTA VICTORIAN

WHAT SHE HAD A MODEST VICTORIAN, WITH COVED CEILINGS AND 100-YEAR-OLD MOLDINGS—AND THE EXPERTISE THAT COMES FROM WORKING WITH HIGH-TECH MATERIALS ON AWARD-WINNING PROJECTS.
WHAT SHE WANTED A HOUSE THAT MAINTAINS ITS DEMURE STREET FRONT—BUT THEN MORPHS INTO A SLEEK GLASS BOX WITH VIEWS FOR MILES.

Anne Fougeron retained her house's antique public facade and ornate stairway (above and left). But the light-filled kitchen (separated from the bathroom by a wall of laminated glass) is anything but quaint (opposite).

Fougeron's living and dining rooms are filled with 20th-century furniture—including Mies van der Rohe's coffee table and lounge chairs (above), and Capellini dining table (opposite). Fougeron designed the bookcases. The modern pieces ease the transition from Victorian to futuristic.

HOW SHE GOT IT Anne Fougeron is an award-winning San Francisco architect known for residential and commercial interiors in which high-tech materials are used to dazzling effect. But it wasn't immediately apparent how she would apply that skill to her own house. After all, the modest 1895 Victorian, on a picturesque San Francisco street, was distinguished by coved ceilings and 100-year-old moldings. "I'm a purist, a modernist, so I wanted rigorous architecture, but I also wanted to convey my love of this house and my admiration for the craft of building," says Fougeron, who was born in France and lives in California with her teenage daughter.

Fougeron's first key decision was to live in the house—for several years—before making big changes. So when she finally decided to start renovating, she knew what was worth keeping. "When the original architecture has merit, I don't believe in slash-and-burn," she says. "I never considered stripping down the interior of my house and turning it into a loft." Among details she decided to preserve are the ornate stairway and vintage wainscoting. Instead, Fougeron focused her efforts on the back of the house (which had great views, and which she could change without alienating the neighbors). The kitchen, Fougeron knew, would be the busiest room in the house. So why not also make it the brightest?

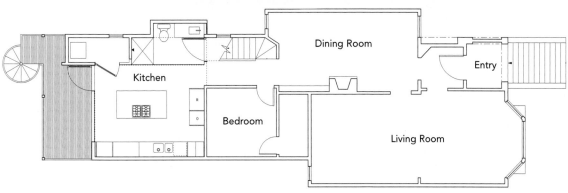

before

Dining Room

Entry

Kitchen

Bedroom

Living Room

Removing the kitchen's old back wall, Fougeron installed a rear facade that is as transparent as a storefront. To achieve maximum openness, she fabricated window frames from thin steel strips, and attached the glass with silicone caulk. Fougeron concedes that had she used energy-efficient windows, with multiple layers, she would have needed thicker window frames.

Inside, the kitchen is from Bulthaup—Fougeron says she couldn't design a better system. Countertops are stainless steel. Fougeron had it buffed with steel wool on a rotary sander for a softer texture. The buffed surface doesn't show scratches (in effect, it's prescratched Fougeron says). The island is Formica, its slickness balanced by the color of the natural cork flooring.

RENOVATION DREAM Next to the kitchen are a small bathroom and laundry room, which Fougeron hid behind a wall of bluish-green glass. The glass is completely opaque, yet it has a reflectivity that makes conventional walls seem lifeless. Upstairs, she created a skylighted master bathroom behind a similar glass wall. Filled with stainless-steel fixtures, "the room needed an organic material," she says, "for balance." So she had two cabinet doors made of onyx, with as much grain as the richest wood. Says the architect, "Contrast is essential for living."

Opposite: Fougeron banished upper cabinets as "incompatible with a busy life." The bar on which she hangs towels and tools is from Bulthaup (which also made her cabinets); Ikea has a comparable system.

What the Pros Know About
Laminated Glass

Nothing says "modern" faster than glass with a pale bluish-green hue. Anne Fougeron says there are several ways to get the effect, including the most basic (painting the back of clear glass). More often, people order sandblasted (also known as frosted) glass. It looks great, but its roughed-up surface scratches and shows fingerprints, Fougeron says. **Her alternative was laminated glass— two layers of glass with a sheet of white plastic between them; the sandwich is permanently sealed. With two smooth outer edges, stains aren't a problem.** Laminated glass costs more than sandblasted, Fougeron says, but not by much.

A master of materials, Fougeron used Formica for her kitchen island, stainless steel for bathroom fixtures and sheets of onyx for a dazzling medicine chest. Her skylight doesn't leak, she jokes, but if it did—with the terrazzo shower floor—"it wouldn't matter."

CONDO MINIMAL

WHAT THEY HAD A TWO-STORY TOWNHOUSE CONDOMINIUM, WITH MORE ROOMS THAN THEIR EMPTY NEST REQUIRED, AND A TRADITIONALLY STYLED EXTERIOR THAT SEEMED TO DEMAND HISTORICAL DETAILS INSIDE.

WHAT THEY WANTED A FLOWING, LOFTLIKE SPACE, WITH LOTS OF STORAGE (INCLUDING A CLOSET AS BIG AS A BEDROOM), WITH FINISHES AND FIXTURES THAT ARE OF-THE-MOMENT—AND DEFINITELY NOT OFF-THE-SHELF.

HOW THEY GOT IT It is one of the most common American building types—the townhouse condominium. It may also be the type that Americans are least likely to renovate: The layout feels preordained. Besides, with condo rules that forbid alterations to the facade, changing the interior seems practically subversive.

But not to David Ling, who turned a three-bedroom, 1,300-square-foot townhouse in a New Jersey suburb into a flowing space more reminiscent of a SoHo loft than the colonial-style house its original details were meant to suggest. Ling, a modernist who has worked for both Richard Meier and I.M. Pei, says, "I blew away as many walls as I could."

For the owners—an empty-nest couple—privacy was less a concern than it might be for a household with children. So among the walls Ling "blew away" was the one that enclosed the master bedroom. That large room, which Ling created from several smaller spaces, now overlooks the double-height living room. The living room in turn opens onto a large kitchen and dining area (which Ling carved out of the spaces originally allocated for those functions, plus an excess downstairs bedroom). Not

Opposite: Behind a facade that gives nothing away, David Ling created a sculptural fireplace of silver-leafed MDF and a "hinge" to highlight the cathedral ceiling.

only does the space feel continuous, but sun from three new skylights now bathes both levels in light. The skylights aren't poky holes, but architecturally scaled openings that help sculpt the overhead space.

Once the rooms were opened up, Ling turned his attention to surfaces, choosing materials that would convey solidity and style. In the kitchen, he had storage cabinets custom-made to support sheets of ⅛-inch steel, which were screwed to the cabinet and drawer fronts. (The steel, Ling says, is hot-rolled—a process that gives it a bluish-black color.) The living room fireplace is also metal—in this case MDF covered with silver leaf, above a hearth of Japanese river stones.

The floors are extra-wide (five-inch) planks of American maple; according to Ling, when the planks are wide, suggesting boards from old-

growth trees, they convey a feeling of solidity. So do the baseboards, which are made of strips of wood faced in more hot-rolled steel (all of which was waxed to resist rust). In contrast to the nearly black steel kitchen cabinets, the cabinets in the upstairs dressing area are white. Ling had them covered in Formica, with strips of steel for punctuation—their syncopation recalling the window walls at Le Corbusier's monastery of La Tourette, or a Mondrian painting.

But the jazziest piece of sculpture in the house may be clinging to the ceiling. Ling hid a crudely shaped peak with white-painted ribs and a black metal "hinge" that draws the eye upward (next page). "I replaced an awkward form," the architect says, "with a form *I* like." Just don't tell the condo association.

Above: In the skylighted closet, steel is used to frame Formica-faced cabinets. But in the kitchen (opposite), steel is everywhere—suggesting a minimalist artwork. The columns, which match the cabinets, were once inside a wall that Ling removed. Arne Jacobsen's curvy chairs soften the composition.

before

What the Pros Know About
Built-in Bathtubs

The urge to customize has moved beyond the kitchen. But unlike refrigerators and dishwashers (which commonly accept false fronts), the typical bathtub presents a face of porcelain. Still, there are ways to achieve a made-to-order look. One is building a tub from scratch—tiling a wood frame covered in waterproof wallboard. But the result is likely to be a "lot more expensive, and a lot less comfortable, than a preformed tub," says architect David Ling. Plus, "it's probably going to leak." A good alternative is to place a low wall directly in front of the tub (with a ledge for shampoo bottles). **Purchase a drop-in tub, which doesn't have a finished front, and build a wooden frame around it.** Add waterproof wallboard and tile. Ling chose hand-cast, one-inch glass tiles from Brazil. The tiles are held together with sheets of brown paper—in front. (Because the tiles are translucent, the normal web backing would show.) But since the paper isn't removed until the tiles are installed, getting the seams to line up isn't easy. Ling's advice: "Hire a very good installer." The lip of the tub can overhang on the tiled box, but in this case, Ling chose to tile right onto the lip. His choice of cobalt blue tiles with light green grout gives the niche an uncommon richness.

RANCHES

BACK

STORY

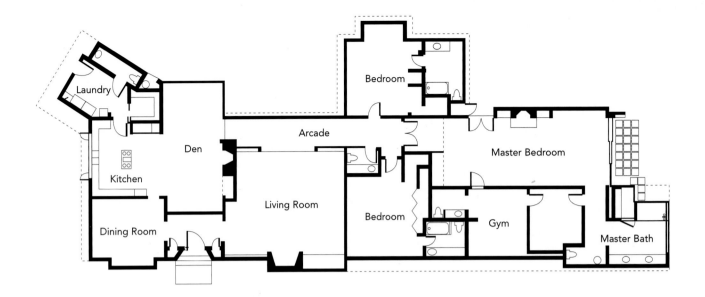

WHAT HE HAD A 1950S RANCH WITH TOO MUCH DARK PANELING (FOR A COLLECTOR OF CONTEMPORARY ART), TOO FEW LINKS TO THE BACKYARD AND A LAYOUT THAT TURNED HIS LIVING ROOM INTO A HALLWAY.

WHAT HE WANTED A PLACE WHERE HIS ART WOULD LOOK GOOD, WHERE HIS YARD WOULD LOOK GOOD (THROUGH GLASS WALLS THAT SLIDE OPEN) AND WHERE HE WOULDN'T HAVE TO PASS THROUGH ONE ROOM TO GET TO ANOTHER.

Opposite: In the dining room, upholstered chairs by Eero Saarinen give a lift to the scissor-leg table, adapted by Tim Clarke from a Jean-Michel Frank design. The Japanese lamp is as spare as Edward Ruscha's art, but restored wainscoting (below) and moldings (above) the painting keep the room from becoming—in the owner's words—"scarily modern."

HOW HE GOT IT Jeremy Zimmer liked his neighborhood—a Los Angeles canyon filled with extravagant flora and modest, mid-century houses. And he liked his property, with several great old oak trees and a surprisingly flat (for a canyon) backyard.

But his house, a 1950s ranch right out of *Ozzie and Harriet*, needed help. There were layout problems: The kitchen and den were at one end, the bedrooms at the other, with the living room as the only connector. That made the house's largest room more of a hallway than a place to relax. And there were decor problems: The interior walls were covered in dark wood, anathema to a collector of moody photos by Sally Mann and paintings by Edward Ruscha. "The wood," he says, "soaked up the art."

Zimmer's goal, he says, was to make the place "modern but not scarily modern." A Hollywood talent agent, he called Tim Clarke, a Hollywood talent just setting out on his own. (Clarke is now one of Los Angeles's top designers.) Clarke had some ideas for enlarging the master

bath and opening the den up to the yard. That required an architect, and the one they brought in—British expatriate Gwynne Pugh—had a few thoughts of his own.

Pugh added an arcade behind the house—providing a straight shot from the den and kitchen to the bedrooms. With sliding glass doors for its entire length, the arcade is both a luminous addition to the living room and a buffer between inside and out. (To strengthen the connection, Pugh and Clarke used the same slate floor in the arcade and on the patio outside.) In the den, there are more sliding panels, including two that form a corner but open away from each other (above)—a detail that Pugh says required "a few engineering shenanigans but nothing too exotic."

Zimmer had ideas of his own. After a trip to Hawaii, he wanted his bedroom and bath to embrace the outdoors. Pugh devised a trio of

bedroom sliders that telescope into a wall, which Zimmer says makes him feel like he's waking up in the jungle, rather than a couple of miles from the freeway. And he created a glass-walled shower that can be entered from the bathroom or the yard.

Zimmer says he had planned a small renovation, but "one thing led to another." (What renovator hasn't said those words?) Still, he respected the simple lines of the original house. The rustic front facade is pretty much as it was, except for the new etched glass and steel front door, which seems to say, "Modernism coming through." In the living and dining rooms, Zimmer and Clarke had to decide whether to keep the wainscoting that covered the lower walls. To their credit, they left it in place, and even replaced pieces where necessary. Otherwise, the rooms could just be new construction, rather than a skillful blend of old and new.

Above: The glass-walled arcade behind the living room (with slate floor and oversize skylight) connects indoors and out. Tim Clarke hung a mirror on the pool house; now it literally reflects the main house. Opposite: In warm weather, Jeremy Zimmer's den goes cornerless.

Left: To avoid having a post in the corner of the den, Gwynne Pugh cantilevered the old roof from new beams. He included an operable window in one of the sliders, for breezes even when the wall is closed.

What the Pros Know About
Resurfacing Fireplaces

Updating a brick or stone fireplace is less difficult than it looks. "People think the outer layer of masonry is an integral part of the fireplace, but it's not," says architect Gwynne Pugh. "If you look inside, you can usually see a gap between the outer layer and the firebox, especially," he jokes, "if you look after an earthquake." That means there's nothing dangerous about a fireplace without its original front. Here, Pugh had the stone chiseled off and replaced by exterior plaster troweled to a smooth texture. **He chose cement-based rather than gypsum-based plaster because it's more fire-resistant and easier to clean.** Removing the old stone helped Pugh get the flush surface he wanted, but it's also possible to leave the masonry in place and plaster over it.

RENOVATION DREAM Designing the spalike master suite that Zimmer asked for—and inserting it into a conventional, wood-framed house—was a task that fell to Pugh. The bedroom's side wall wasn't load bearing, so Pugh (who is also an engineer) was able to honor Zimmer's request for nothing but a wall of glass between bedroom and garden. But Pugh went further, dividing the wall into three glass sliders that can disappear entirely. The trick, Pugh says, is that instead of trying to hollow out the existing wall to create a pocket, he simply added a pocket outside the existing wall, then covered it in siding. It helps, Pugh adds, to have a good door manufacturer; he used Fleetwood

(fleetwoodusa.com). In the bathroom, Pugh added a Kohler tub in a new glass compartment. To create the frameless corner, he butt-jointed two pieces of tempered glass. But since the glass walls can't support the ceiling, Pugh hid a steel pipe column in the wall to the right of the tub, as close to the corner as he could get, and cantilevered the ceiling from the column. (If the ceiling settled even a little, it would crack the glass.) He also had to leave room for plumbing (for the shower) in the same wall that the bedroom wall slides into. Technologically, none of what he did was difficult, Pugh says, "The trick was working out every last detail in advance."

LOST
AND
FOUND

WHAT THEY HAD AN ODDLY CONSTRUCTED POST-AND-BEAM HOUSE IN AUSTIN, TEXAS, WITH POTENTIAL ONLY THEY COULD SEE. **WHAT THEY WANTED** A KIDPROOF HOME WITH MATERIALS AND FIXTURES THAT REFLECT THE OWNERS' "BACK TO CAMP" AESTHETIC—AND LOVE OF RECYCLING.

A quirky house requires quirky furniture. Opposite: Seating is from a filling station waiting room; the new rosewood parsons table sits beneath an Ingo Maurer chandelier. Tiny, outdoor-style spotlights ensure that ceiling beams remain on view at night. Above: A discarded industrial prep table became the kitchen island.

HOW THEY GOT IT There's no denying that the house Kimberly Renner bought was odd. "I didn't really know what she saw in it," says Mell Lawrence, the architect Kimberly and her husband, Dan, hired to make the building livable. Fern Santini, the couple's interior designer, says, "I'm a pretty good judge of what's possible, but I didn't see it here." But Kimberly, a former director of Austin's recycling program, prides herself in turning trash to triumph. "She's a junker of the highest order," says Santini.

The house would be Kimberly's greatest challenge. The building, in thick woods outside of Austin, was the creation of a weekend builder. He imitated the post-and-beam construction he saw in magazines, but he made his posts and beams out of two-by-sixes nailed together—presumably, Dan says, because, working alone, he couldn't lift real beams. As a result, the house had a crafty, old-world quality. Says Lawrence, "It was built in the 1980s, but it looked more 1880s." He isn't being kind.

And yet many of the "posts" and "beams" had been covered in Sheetrock. Playing to the house's strengths, Lawrence stripped off as much of the Sheetrock as he could, making the interior more barnlike. And then he hid the remaining Sheetrock behind pine boards laid horizontally. Channeling Kimberly's spirit, Lawrence says, he didn't buy the

The foyer wall—which once reached the ceiling—was cut down to six feet, exposing pipes that lead to the second-floor bathroom. What remained of the wall was covered in steel sheets.

before

What the Pros Know About
Cold-Rolled Steel

Says Kimberly Renner, "We reduced what was a floor-to-ceiling wall opposite the front door by several feet, making it more of a partition over which light could flood into a living area." But no ordinary wall would do for her. So Lawrence sheathed the partition in sheets of quarter-inch-thick cold-rolled steel, available at any metal fabricator. **The sheets, which aren't structural, were simply bolted to studs.** Then the material was waxed and buffed, which, Kimberly says, "gives it a deep, marbled luminescence." This nontraditional material—at least in a residential setting—lends an industrial elegance. "Guests love to touch it," says Kimberly.

What the Pros Know About
Barn Doors

Increasingly popular with people who have never owned a barn, these sliders have an advantage over pocket doors: They don't have to fit into hollow walls. That means less construction, and allows for thicker doors. Kimberly's door is made of galvanized metal, but almost any material will work. Barn-door hardware, which must be hung from studs, can cost less than $100, or more than $1,000. The track can be hidden behind a valence, or left exposed (as in the Renners' bedroom). **Good hardware is designed to help level the door after installation, for a perfect slide.** One problem: There's no way to hang art on the wall the door slides past. San Diego architect Jennifer Luce has a solution: "Hang art on the door," she advises.

best pine, just medium grade. And, he said, "we let it shrink for a month or so before we painted it. That caused a mix of tight gapping and wide gapping, which is part of the aesthetic." Most of the pine was covered with two coats of white enamel, creating a neutral backdrop for the objects Kimberly brought into the house, including chairs from a filling station waiting room. In the kitchen, Lawrence covered the knotty pine cabinets (previously shellacked) in white paint; the white, he says, is a unifying element of the design.

When the renovation began, the couple didn't have children. But while construction was underway, two sons (now three and four) were born. "Although the decision to make walls from pine was design-based," Kimberly says, "I now understand the value of a kidproof material. Toys can't dent it and neither spaghetti nor crayons can stain it."

Perfection wasn't the goal. When Lawrence removed part of a wall separating the living room and foyer, pipes from the upstairs bathroom were suddenly exposed. There was no way to move them; instead, he covered them with galvanized-metal sleeves, which complement the

"paneling" of cold-rolled steel (see box, page 69). "We slipcovered them," he says. And everybody knows slipcovering saves money.

But that doesn't mean the renovation was slapdash. When you renovate, Kimberly says, "you end up doing something to every surface in the house." The couple also built an addition containing a new master bedroom and bath, which is about as offbeat as the old construction. The door to the bedroom is corrugated metal; the floor is waxed concrete; and the ceiling is canvas, a kind of indoor tent designed by Kimberly, made by a local sailmaker and hung by Dan. In the bathroom, Kimberly hung an ornate gilt mirror, but for faucets she chose "the things you attach hoses to outside a house." Says Santini, "She had the self-confidence to know that this renovation was going to work."

RENOVATION CHALLENGE "In the master bath, the valves I connected to my antique brass spigot were intended for outdoor use and not made with the integrity needed to prevent leaks. So the faucet always drips," says Kimberly. "The good news is that our cat, Lizzie, uses the puddle it creates in the tub as her personal watering hole."

The master bathroom is new—but it reminds Kimberly of the summer camps she attended (apparently a very happy camper!) as a child. Industrial sinks (opposite, right) are paired with metal hospital cabinets obtained when a local air force base shut down. Previous spread (bottom right): In the master bedroom, an old movie-theater-poster frame contains a mirror.

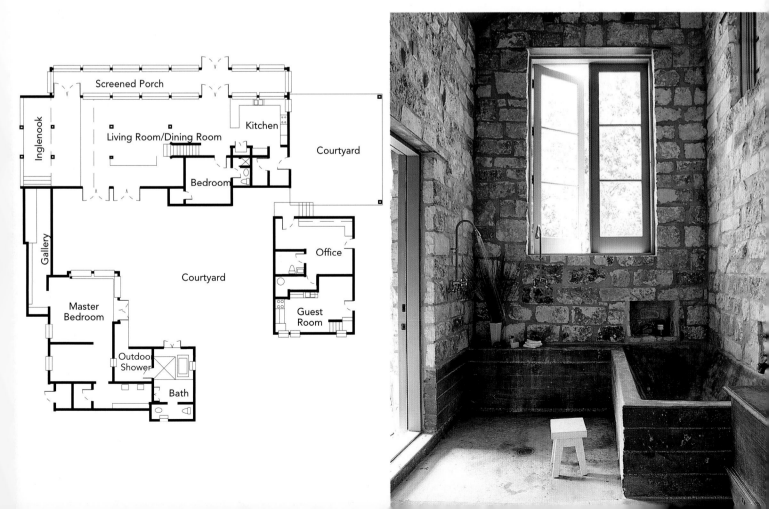

Screened Porch

Inglenook

Living Room/Dining Room

Kitchen

Courtyard

Bedroom

Gallery

Office

Courtyard

Guest Room

Master Bedroom

Outdoor Shower

Bath

GOOD AS GOLD

WHAT THEY HAD A 1950S RANCH IN HIGH POINT, NORTH CAROLINA (HOME TO THE NATION'S LARGEST FURNITURE MARKET), WITH A WOODED LOT AND GREAT POTENTIAL FOR CREATING BOTH PRIVATE AND PUBLIC SPACES.

WHAT THEY WANTED A HOUSE IN WHICH A FURNITURE MOGUL CAN ENJOY QUIET DOMESTICITY—AND STILL ENTERTAIN CUSTOMERS BY THE HUNDREDS.

before

After lightening the house with white paint, Mitchell Gold and Bob Williams didn't need old skylights (above). "They leaked," says Williams. "It was easier to remove them than replace them."

What the Pros Know About
Pocket Doors

Pocket doors are beloved by architects—because furniture can be arranged without regard to door swings. Pocket doors can range from narrow bathroom or closet doors to giant room-dividing panels, and can be made of almost any material. What they have in common is that they disappear into seemingly solid walls. Some small pocket doors come in kits, with the pocket ready to be hidden behind Sheetrock. **For the largest doors, a special, extra-thick wall may be required. But a medium-size door can be accommodated by turning studs parallel to the wall, creating a frame with an opening down the middle.** Of course, there won't be room in the wall for pipes and ducts, and it's important to plan carefully if light switches and outlets are required. Even well-installed pocket doors are harder to close than swinging doors. Says Bob Williams, "I wouldn't want a pocket door if I had to open and close it all the time—they can be difficult. But for doors you only use occasionally, they're great."

HOW THEY GOT IT "We had 90 people for dinner the other night," says Bob Williams, "and it worked out great." The setting was the house that Williams renovated for himself and Mitchell Gold, of the furniture company that bears his name. Twice a year, in April and October, Williams (the company's creative director), Gold and tens of thousands of manufacturers and buyers descend on High Point, North Carolina, for the International Home Furnishings Market. With hotels oversubscribed, many attendees rent houses from High Point residents. Gold rented for 25 years, always living with other people's decor, he says. But now Gold and Williams have a house to themselves, a 1950s ranch that Williams made both cozy enough for a tête-à-tête and grand enough for corporate entertaining.

Williams had done renovations for the company before but always in commercial buildings. "A house is more of a challenge," he says, "because you wake up here, so if something isn't right, it's going to annoy you," he says. Luckily, the house he and Gold bought has a great site—a woody 4½ acres—and a generous 3,500-square-foot interior, which the men augmented with indoor and outdoor entertaining spaces. Most of the walls were covered with shiplap siding. "It was a natural pine, and after 50 years, it had become kind of dull-looking," says Williams. So after adding more of the siding, to strengthen the connections between old

Above: Gold's corporate chef prepares banquets in the house's newly enlarged kitchen. Come morning, what had been a cooking island serves as a homey breakfast table.

Dining Room
(for larger
functions)

Outdoor
Patio

Formal
Dining
Room

Living Room

Bedroom

Master Bedroom

Garage

Kitchen

Bedroom

Bedroom

Opposite: "Part of me wants more and more color," says Williams, who let loose in a guest bathroom (top left) with tiles from a local home-improvement center. In the master bedroom (bottom right), furnishings adhere to a more neutral palette. The brushed-metal four-poster bed echoes the new, bronze-colored windows.

and new, he painted the boards white. "That really lightened up the house," says Williams. The shiplap still has lots of texture, but it also provides a neutral background for furniture (including many Mitchell Gold pieces) and black-and-white photographs of the rural South.

Because they often entertain here, Gold and Williams wanted an easy flow between the public spaces. That required a wholesale reorganization: The family room became the kitchen, the old kitchen became the formal dining room and the screened back porch became a larger dining room for parties. But since this is also a home, the men needed private spaces. The solution was to install pocket doors wherever they could, creating a layering of personal and public zones. They also changed the house's windows: Creaky wooden frames gave way to bronze-toned metal casements, which, Williams says, make the house feel "more substantial."

In the renovated bathrooms and new kitchen, counters are Avonite, a solid surfacing material made from acrylic mixed with bits of stone. "The counters look like terrazzo, and they wear incredibly well," says Williams. Avonite is sold in one thickness ($\frac{1}{2}$ inch), but sheets of the solid surfacing (like those of Corian, a competing product) can be melted together, a process that leaves no visible seams. That allowed Williams to create a vanity top that appears to be five inches thick. Despite Williams's ingenuity, Gold sighs, "We spent more on the makeover than we did on the house." But he also knows it was worth it.

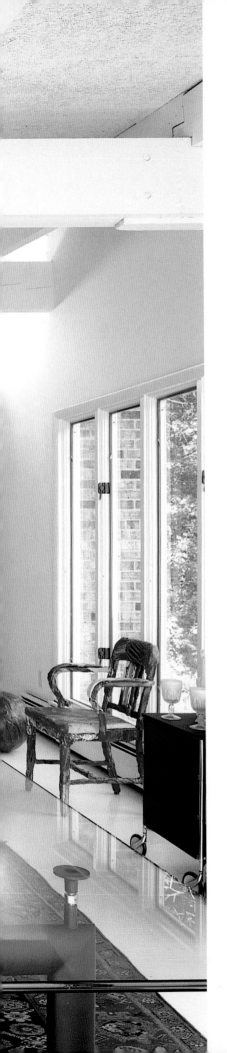

CAPITAL IMPROVEMENT

WHAT THEY HAD A HOUSE TURNED DOCTOR'S OFFICE, WITH A BIZARRE LAYOUT AND A SURPRISING LACK OF SUNSHINE GIVEN ITS GENEROUS EXPANSES OF GLASS.

WHAT THEY WANTED UNOBSTRUCTED LIGHT AND VIEWS—WITH WHITE PAINT COVERING ANYTHING THEY DIDN'T LIKE BUT COULDN'T AFFORD TO REPLACE.

HOW THEY GOT IT Sam Gilliam, a leading African-American painter, used to cover the walls of his homes with artworks. But these days, there are only two canvases in the house he shares with art dealer Annie Gawlak. The reason, Gilliam says, is simple: "Now we have light."

The couple's house overlooks Washington D.C.'s Rock Creek Park and offers high ceilings and exposed beams more typical of California than the capital (with its penchant for Federal-style architecture). But the previous owner, a doctor, had used it as an office—which is why, Gawlak says, "it had the worst house layout I had ever seen. There was no rhyme or reason." Adds Gilliam, "For a house with walls of glass, it was incredibly dark."

In fact, the place, divided by endless partitions, was so depressing, Gawlak says, that it sat on the market for three years. "Since we were the only people who had expressed any interest in it whatsoever, the seller kept trying to sweeten the deal. But we knew it would need work, and we had no cash." Ultimately, the couple obtained a mortgage, with $20,000 "extra" to be used for renovation. "But $20,000 doesn't buy a lot of renovation," Gawlak says.

It does buy a fair amount of demolition, so the couple began by removing everything that stood between them and the views and light

Sam Gilliam's contributions to the house he shares with gallerist Annie Gawlak include two colorful canvases and a fabric-wrapped column (previous spread). Walls, floors and ceilings are covered with a Durham paint aptly named High Hiding White. In one part of the foyer (opposite top), Gilliam and Gawlak hung copper-painted washers, from a local hardware store, in geometric patterns.

What the Pros Know About
Skylights

Chip Bohl, a Maryland architect, says that the reputation of skylights for leaking may no longer be deserved. Certainly, old skylights, with panes of glass set into metal frames, had "weatherproofing problems," he admits. **And skylights that open are more likely to leak than ones that don't.** Still, Bohl says he has had great results with operable skylights, including those from Velux, a European company that offers such features as electric motors, moisture sensors that close skylights automatically at the first sign of rain and blinds to control light and heat gain (velux.com). Bohl says he doesn't always choose the biggest skylight his clients can afford, preferring several smaller ones as a kind of accent lighting. "A skylight in a corner," he says, "brings light to the darkest part of the room, so you get the maximum benefit from the sun." It's rare that a skylight is intended to capture a view, but even so, every skylight should be cleanable. "Some operable skylights flip around so that you can wipe them down from the inside," Bohl says. Because warm, moist air naturally rises, condensation tends to form on skylights, so it's important to look for insulated models with condensation traps.

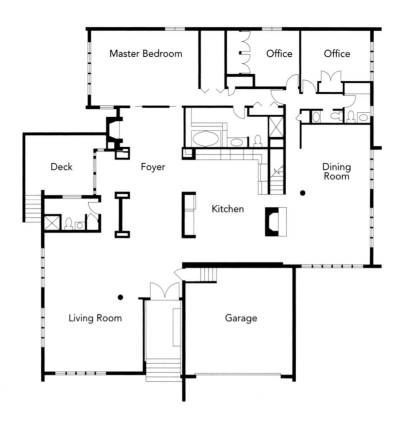

Master Bedroom Office Office

Deck Foyer Dining Room

Kitchen

Living Room Garage

they craved—even when that meant tearing out ductwork for the house's central air-conditioning (no luxury in Washington, D.C.). Eventually, when they had the money, they installed a new system, this time with ductwork in the basement. Designer Mary Douglas Drysdale helped them streamline the layout to create what she calls "a liquid feeling of movement."

Mostly, the couple renovated with paint—nearly all of it white—which covers not only walls and ceilings, but cabinets, floors and exposed beams. The kitchen floor, made of brown tiles, was coated in white latex deck paint, which, Gawlak says, forms a "surprisingly durable surface." She adds, "As an artist, Sam believes in the power of paint to solve all problems."

In the foyer, the couple peeled off old carpet, exposing the house's plywood subfloor. But since they couldn't afford new flooring, they had to find a way to make the subfloor more appealing. Barbara Billet, a local artisan, painted the plywood white. When the white paint dried, she added a layer of beige. Before the beige paint dried, she took a squeegee with sections cut out (it resembled a comb) and ran it over the plywood, creating streaks of white and beige. The effect is "very ethnic, like an African textile," says Gawlak. Billet did one piece of plywood at a time. That way "each sheet has its own rhythm, its own beginning and end," says Gawlak. "We didn't try to mask the fact that there were individual plywood planks."

Really, the couple didn't try to mask anything. "You can look to the left and to the right, and up and down, and see the park," Gawlak enthuses of the newly open floor plan. Gilliam says the views of nature inspire him, especially the waning days of autumn, "when the leaves fall and everything becomes transparent." Says Gawlak, "We're still living with exposed subfloors and a 1950s kitchen. But now it's a wonderful house."

LOFTS

WITHOUT BORDERS

WHAT THEY HAD A CAVERNOUS LOFT IN AN INDUSTRIAL BUILDING IN CHICAGO, PLUS THE MODERNIST CONVICTION THAT IN ARCHI-TECTURE "LESS IS MORE."

WHAT THEY WANTED A HOME WITH AS FEW WALLS AS POSSIBLE, SMOOTH EXPANSES OF CONCRETE AND LOTS OF STORAGE SPACE FOR ART THAT ISN'T ON DISPLAY.

Opposite: The kitchen's range and fridge are bracketed by no-pull, custom-made stainless-steel cabinets (reaching all the way to the ceiling). Chrome-and-leather stools are by Le Corbusier.

HOW THEY GOT IT Most renovations start with carting things away—filling dumpsters with old fixtures, walls and cabinets. For Chicagoans Neil Frankel and Cindy Coleman, that's where their renovation began—and ended. The couple's goal was to keep their loft (in a red-brick turn-of-the-twentieth-century building in the historic Printer's Row district) as open as a commercial space. So they designed their home without a single interior wall, except the ones around the bathrooms.

But that doesn't mean the renovation was easy. Mies van der Rohe's pronouncement "less is more" has long been understood as "less costs more." Coleman and Frankel agree with both statements. After all, if you want to leave pipes and ductwork exposed, they have to be perfect. And running wires through the apartment, when there aren't any interior walls, means burying them in the floor; the floor then has to be refinished. But with electric outlets underfoot, furniture can be placed midroom, rather than against the walls, as in conventional apartments. Overhead, track lighting is ready to follow the furniture wherever it ends up.

Though wall-less, the 3,500-square-foot space feels like it's divided into zones. Several sections of the floor are raised about one foot. Guests, Frankel observes, like to stay in the full-height spaces when they're standing, and then move to the platforms when they're ready to sit. "We never tell them to do that," he says, "it just happens."

Furnishing a wall-less space means that pieces have to look good from every angle. Since TVs aren't always pretty from behind, Frankel designed a stand (with perforated metal back) that swivels. The couple

also wanted their fireplace to be just as sculptural. "We could have had a flue in the back of the building," says Coleman, "but when we began researching fireplaces, we started to realize that flueless gas fireplaces are better—not just because they're easier to install and maintain, but because gas burns much cleaner than wood." (See box, page 96.) The couple avoided ceramic log substitutes. "The focus," Coleman explains, "should be on the flame."

The couple's nine-year-old daughter, Emanuella, has graduated from a stainless-steel crib to a powder-coated aluminum loft bed; nearby is a chaise designed by Le Corbusier, which her parents thoughtfully commissioned at $5/8$ scale. It's never too early to gather modernist credentials. **RENOVATION DREAM** Coleman and Frankel don't like having more than a few artworks on display at any one time (they believe that it's easier to appreciate objects in isolation). So when they were building closets (which define, without enclosing, the loft's bedrooms), they went all the way to the ceiling. The upper cabinets provide perfect long-term storage for art between showings.

Opposite (top): The master bedroom sits between the fireplace and Mies van der Rohe's minimalist chaise. A pivoting glass panel hides the master bathroom. Bottom row: To the daughter of modernists, *a* is for Aalto, *z* is for zigzag: Emanuella's round table and chairs are Alvar Aalto classics; the "Zig Zag" chair is by Dutch master Gerrit Rietveld. The black print is by Richard Serra.

What the Pros Know About
Gas Fireplaces

Typically, gas fireplaces require double-walled ducting (a small pipe within a larger pipe that brings in air for combustion and vents carbon monoxide out). This ducting can either fit into an existing chimney (or metal stovepipe) or vent horizontally via an outside wall. But to achieve maximum openness in their loft, Cindy Coleman and Neil Frankel chose a ventless fireplace, which uses room air for combustion. A typical setup consists of a "gas log set" (the "logs" are concrete or ceramic) and a metal firebox. **Options include thermostats, remote controls or wall switches and forced-air blowers.** Most models come with sensors to warn of oxygen depletion (a danger that explains why some municipalities don't permit ventless fireplaces in bedrooms). Majestic (majesticproducts.com) manufactures ventless propane and natural-gas fireplaces, including see-through (two-sided) and peninsula (three-sided) models. All require gas lines, and most require electrical wiring as well. Kyle Smoot of Dixie Products (a large fireplace dealer in Roanoke, Virginia) says many customers spend less than $500 on their ventless units (though prices can hit $2,000 or more). Smoot says ventless fireplaces "are not meant to run for hours on end—combustion, no matter how efficient, produces moisture, and you can have mold or mildew problems if you overdo it." Ventless fireplaces can be outfitted with metal fire screens, but never glass doors—which could cause dangerous heat buildup.

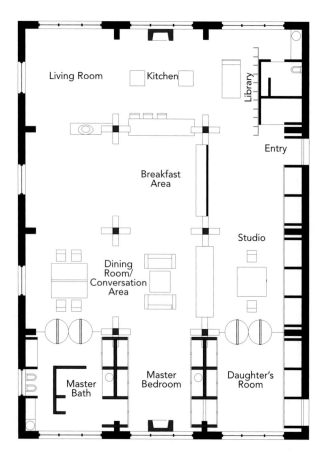

What the Pros Know About
Concrete Floors

Getting an exposed concrete floor to look good requires "continuous supervision," the couple learned. In their case, the contractor had only one person troweling 3,000 square feet. That resulted in uneven distribution of the material, which led to more cracking than usual, says Coleman. Also contributing to cracking: the contractor's failure to extend expansion joints all the way to the columns. But even perfectly installed concrete floors can develop cracks—in the worst places. One of Frankel's engineering professors used to say, "Concrete doesn't go to college; it doesn't know where it's supposed to crack." But it can go to finishing school: **The couple's new floor was shaved, polished and coated with a chemical sealer, for a smooth-to-the-touch surface.**

DRAWING ROOM

WHAT THEY HAD A RAW LOFT SPACE WITH COLUMNS IN THE MIDDLE OF THEIR FUTURE LIVING ROOM, A FIREPLACE THAT NEEDED TO BE MOVED AND A KNACK FOR RENOVATING ANTIQUE HOUSES IN NEW ENGLAND.

WHAT THEY WANTED A STYLISH HOME THAT WOULD FUNCTION AS A DAD'S PIED-À-TERRE, A SON'S BACHELOR APARTMENT AND A MEETING SPACE FOR THEIR BURGEONING BUSINESS.

HOW THEY GOT IT When Marc Brown was a struggling cartoonist raising a family outside Boston, he bought old houses, fixed them up and sold them at a profit. Not only did the projects supplement his income, Brown says, but "renovating has allowed us to live in some pretty wonderful places—they were in such bad shape that no one else would have them." Then Brown's character, Arthur the Aardvark,

became the basis of a PBS series, and an empire was born. Brown is still renovating, but on a slightly grander scale.

Among his recent projects is a New York City loft, a home for his twenty-something son, Tolon (who handles the licensing business). "Tolon shares the same genetic disease of wanting to fix up properties," says Marc. So father and son spent a year looking for a place they could redo together. The 3,700-square-foot loft they settled on was "just cavernous, and a little bit daunting," says Marc. So he brought in the A-team: architect Deborah Berke, who is celebrated but, Marc says, "totally non-diva," and interior designer Thomas O'Brien of New York's Aero Studios.

Marc asked them to play to the place's strengths. "It's a loft," he says, "and we didn't want it to feel like some Upper East Side apartment." Large concrete columns dominated the space. Berke had designed

Above: The Browns' loft is at once clubby and spare. The dining table, by Jean Nouvel, is appropriately named "Less." Aero's upholstered furniture could be "More." A painting by Terry Turrell hangs from a recessed picture rail over a B&B Italia credenza.

Storage

Master
Bedroom

Guest
Room

Office

Media Room

Living Room

before

What the Pros Know About
Translucent Acrylic

Translucent panels are a way of dividing spaces while giving light free reign. For applications where glass would be impractical (one sliding door in this loft is eight by ten feet) many designers use synthetic materials, including Lumasite (made by the American Acrylic company on Long Island, americanacrylic.com). Lumasite comes in dozens of thicknesses, colors and degrees of transparency—one of the most popular resembles the paper used for shoji screens. (It ranges from about $2 to more than $10 per square foot, depending on order size and other factors.) **The material is easy to cut and can be screwed to metal or wood door frames,** but in this case architect Berke didn't want visible fasteners. "The idea," says project architect Robert Schultz, "was to keep the doors looking fairly abstract." That meant gluing the panels to their custom-made aluminum frames, a process that required experimentation. Even now, "if you look closely, you can see some of the glue lines," Schultz says. "But because it's only on the frame, and not where the light passes through the doors, it's hardly visible." The doors are hung on tracks from the German company Häfele (hafeleonline.com), which makes high-end hardware so good-looking, you won't want to hide it.

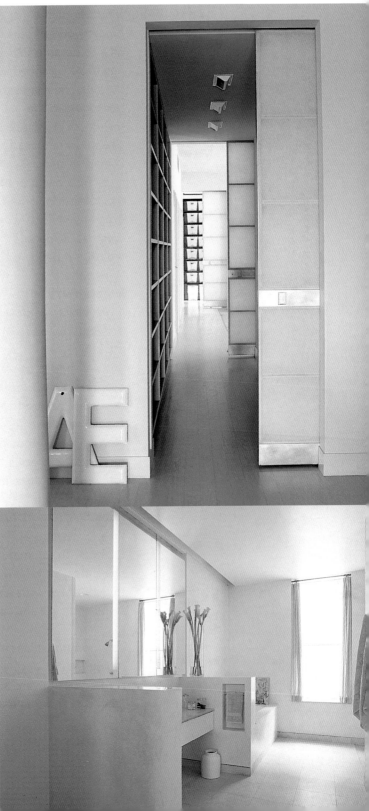

another loft in the same building, "where we hid those very columns." But this time she left them exposed. After all, she says, "Marc has had a life-time connection to old buildings."

Just off the elevator, one of those columns turns visitors toward the windows at the far end of the loft. "Successful interiors allow you to project yourself beyond the four walls," Marc says, adding, "It's especially nice to be able to see outside just as you're coming in." Still, openness had to be balanced with a need for privacy. (When Marc is in New York on business, he shares the loft with Tolon.) The long passageway linking the bedrooms to the main living area can be closed off at three points. But the doors disappear into pockets when they aren't needed. And even when they're closed, they let light pass through: The doors were made of Lumasite, an acrylic that resembles rice paper (see box, page 103).

To allow O'Brien to create a comfortable furniture arrangement in the living room, Berke moved the fireplace several feet. (According to project architect Robert Schultz, because metal flues can be angled as much as 30 degrees, it's surprisingly easy to move a fireplace, even in a multistory building.) Berke designed the marble fireplace surround. Where there would normally be a mantel, there's a far simpler detail—a metal insert that separates the upper walls, which are Sheetrock, from the lower, which are polished plaster.

But the place isn't slick. Ceilings are a gypsum-based plaster called Structolite, which is normally used for base coats. Here, it was left unpainted to suggest structural concrete. Hiding the conduits leading to overhead light fixtures would have required lowering the ceiling, so the designers decided to leave the conduits exposed. "That's no problem, except you have to carefully plan everything in advance," Schultz says. New floors are four-inch slats of red oak. Before they were painted, the paint was thinned, so that the grain shows through. In the kitchen, with tongue-and-groove maple walls, a pot rack echoes the picture rail that ties the loft together.

RENOVATION DREAM Marc and Tolon wanted to be able to rotate their contemporary art collection without damaging the walls. Schultz located an aluminum picture rail, designed to be inserted into Sheetrock, from New York's Milgo/Bufkin (milgo-bufkin.com). According to Schultz, you cut a slit out of the Sheetrock, screw the channel to the studs and tape and spackle over its flanges—in the end all you see is the rail itself. The result is a practical hanging system that also provides a common trim element from room to room—a modernist equivalent of a molding. And, according to Berke, there's a bonus: "By dividing the wall plane, you get the sense that there's that much more to it," says the architect, and that makes the ceilings seem higher.

Opposite: In Tolon Brown's bedroom, vintage movie posters hang over a B&B Italia sofa. A storeroom contains plastic tubs, "like the kind the Post Office uses, without the writing on them," says Tolon's father, Marc Brown.

BOSTON BEACON

WHAT THEY HAD A RAW SPACE IN A NEW LOFT BUILDING, WITH 16-FOOT-HIGH WINDOWS THAT WERE STRIKING—BUT ADMITTED TOO MUCH HEAT AND LIGHT.
WHAT THEY WANTED A DUPLEX APARTMENT THAT TAKES FULL ADVANTAGE OF THE WINDOWS BUT CAN BE AS COZY AS THE VICTORIAN HOUSE THEY USED TO LIVE IN.

David Hacin designed this Boston loft building, and then his own apartment, with its fifth-floor terrace (above). Combining ceiling fans with air-conditioning as Hacin did (opposite) uses less energy than air-conditioning alone: With efficient fans creating cooling breezes, it's possible to use less air-conditioning without sacrificing comfort.

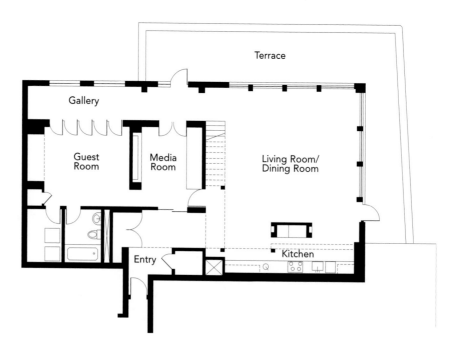

The floor plan shows the layout with the following labeled areas: Terrace, Gallery, Guest Room, Media Room, Living Room/Dining Room, Entry, and Kitchen.

HOW THEY GOT IT When David Hacin, a Princeton- and Harvard-trained architect, began designing a new loft building with 99 units in Boston's gentrifying South End, he had no idea that he would end up living in the building. But then the developer offered him one of the best units—a 25-by-50-foot corner space with 20-foot-high ceilings. Hacin (with partner Tim Grafft, who helps market Massachusetts to foreign tourists) decided to buy the loft. Its most notable feature, 16-foot-high windows, was both an opportunity and a challenge, Hacin recalls, "The best thing is that the space is flooded with light. But because the windows face south and west, I knew we would have tremendous light-control and heat-gain issues." (See box, page 112.)

The couple had been living in an old Victorian, and Grafft loved the coziness of its rooms. So Hacin resolved to create enclosed private spaces within the loft—without blocking any of the dramatic double-height windows. His solution was to design a "box"—a two story building separated from the windows by about five feet. It contains a guest room on the first floor and a master bedroom on the second. On both levels, pivoting shutters control light and privacy. The shutters are solid-core doors that Hacin coated in a high-gloss, oil-based paint and mounted on stock hardware. When they're open, the interior rooms share in the light and views. "As the sun crosses the sky," Hacin reports, "the shutters throw fabulous shadows into the rooms. And then, at night, you can close them all, and have complete darkness inside."

Hacin also covered the windows in drapes of a synthetic linenlike fabric. "They really soften the space, visually and acoustically," Hacin says, "and when light filters through them, there's an ethereal quality to the space." But most of the time, the drapes are open. "The living room is the

connection to the sky and the view," he explains. "And the box is all about making the place feel cozy."

RENOVATION CHALLENGE Hacin says his one regret was choosing black slate for his foyer and gallery. "Black floors look great, but they are very, very difficult to maintain," says Hacin—explaining that dirt, scratches and general wear and tear all take their toll. When Hacin was growing up, his father (also an architect) used black floors in his house, "and my step-mother complained about it. Now I understand," says Hacin. But he wouldn't go white, either "If I was picking another floor, I would pick something midrange," the architect advises.

Above: A five-foot-wide gallery separates the bedrooms from the loft's double height windows. Opposite: A pivoting panel allows the guest room to partake of light and views via the dining area.

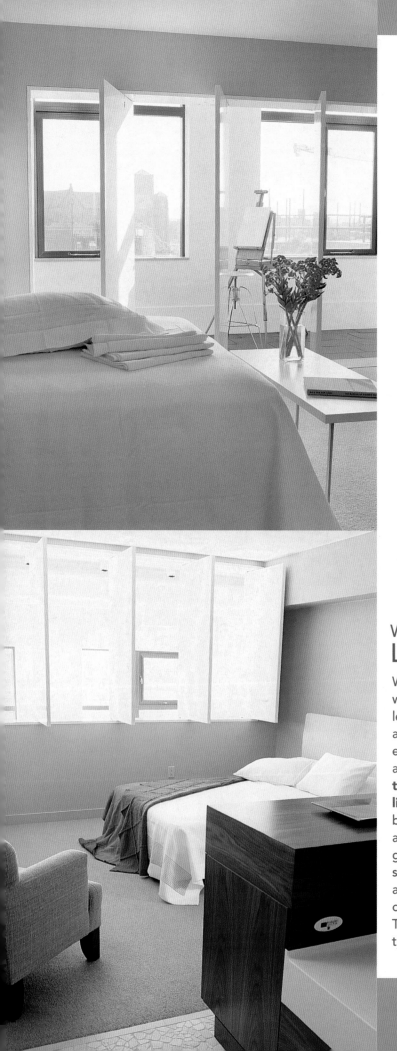

What the Pros Know About
Light-Filtering Film

When he designed the building, most of which faces west, David Hacin knew that sunlight would be a problem. So he specified UV-filtering windows (made with argon gas between the layers). But even that wasn't enough to keep the afternoon sun from turning his apartment into a greenhouse. **So he coated the floor-to-ceiling windows with Solis VK-70, a top-of-the-line polyester film impregnated with silver.** The film blocks UV rays and 55 percent of heat transmission, according to its manufacturer. Though it has a slight greenish-blue tint, Hacin says it's clear enough to satisfy his architect's eye. Overall, he says, it's a big advance over earlier films, which would bubble, streak or darken over time—"leaving you with a disaster." The cost is $12 to $18 per square foot installed for typical residential uses (chbwindowfilm.com).

If you don't mind the lack of privacy, Hacin says, a bathroom open to the bedroom is the way to go, because combining the rooms makes both of them feel bigger. To give bathroom fixtures just inches from his bed the look of furniture, he placed the sink and tub on walnut bases. Each has a Corian "collar," so that spills won't ruin the wood.

LIGHT
INDUSTRY

WHAT THEY HAD A DOWNTOWN LOFT IN A CITY WHERE DETACHED HOUSES ON VERDANT HILLSIDES ARE THE NORM. WHAT THEY WANTED AN INDUSTRIAL-STRENGTH HOME WHERE DETAILS ARE SO BEAUTIFULLY CRAFTED, ALL THE VIEWS THEY NEED ARE INSIDE.

The owners' spirit of adventure means a baby grand piano (top right) coexists with Snap-on Tool chests (used in place of kitchen cabinets) and a flat-screen TV (top left) hangs from chains. Some pieces, like the restaurant kitchen sprayer, are off-the-shelf (bottom left); others, like the syncopated stairway (bottom right), by Savannah College of Art and Design professor John Pierson, are anything but.

HOW THEY GOT IT "In San Diego, everyone dreams of suburban living," says architect Jennifer Luce. So the developers of a loft building, unsure of the demand in Southern California for raw residential spaces, had a brilliant marketing idea: They offered potential buyers a free consultation with their architect. That's how Luce met Tom Felkner and Bob Lehman. "She said, 'Let me give you a few ideas,'" remembers Felkner. Two years later, Luce and partner Sharon Stampfer had designed every aspect of the couple's loft, including most of the furniture. The job not only took longer, it also cost more money than the clients ever expected. "I think of it as investing in a work of art," says Felkner.

Felkner and Lehman had moved from Chicago to San Diego to begin a new life. (The two men, who both toiled in the corporate world, now own Bourbon Street, a jazz and blues club.) And so they weren't timid when it came to renovation. Luce says that, of all the alternatives she showed them, they chose the one that allowed maximum openness. That meant enlarging the hole in the concrete slab that separated the two levels—to make room for an especially dramatic stairway. The stairway sits on a concrete beam sunk eight feet into the floor (it helps to live in a building designed to hold industrial equipment). But wanting the space opened up isn't the same as wanting it stripped down. Each step is different, a cherry-wood sculpture created by Georgia artist John Pierson.

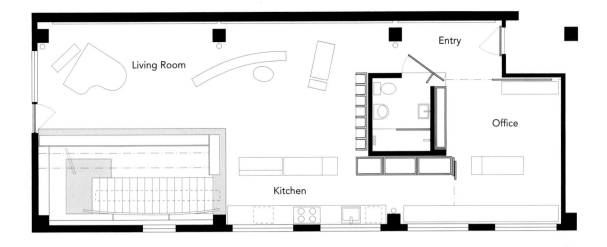

Living Room

Entry

Office

Kitchen

To maintain the open feeling of the loft, Luce grouped fixtures in an elegant "service module." The upstairs bathroom is made entirely of stainless steel—ceiling and floor included. The downstairs bathroom, by contrast, "is all about voyeurism," Luce observes. The glass is layered with Japanese parchment paper, with the number of layers of paper determining the opacity of the glass. (The pane in front of the toilet is the most opaque; the pane in front of the sink, the least.)

The bedroom itself was going to have walls of poured concrete, but then Pierson created cherry panels with holes that accept brackets for custom-made shelves. "The cherry provided a level of warmth—you need to feel nurtured in the sleeping area," explains Luce. "The details keep the loft from seeming too minimal," says Felkner. "With so many rich details, friends who visit always notice something new."

RENOVATION DREAM It's hard to know where to store small things in a big loft. But in a New York City boutique, Felkner and Lehman saw a galvanized metal box that they thought would come in handy for everything from antique toys (Lehman collects them) to rarely used dishes. "But it was around $100," Felkner says, "and we needed something like 100 of them." Then he noticed that the manufacturer's 800 number was stamped in the metal. "We called and were able to order the same boxes for $18 each." Luce designed a cherry frame with stainless-steel trim, and the men—who were thinking there would be doors in front of the boxes—decided that they liked the exposed grid. (See opening spread.) Silicone tape on the bottom of the boxes lets them slide out easily. Since there are no labels on the boxes, Felkner used his computer spreadsheet program to make a key to their contents.

The downstairs bathroom, made of glass, is "a jewel box," says architect Jennifer Luce. The upstairs bathroom (opposite) is encased in stainless steel. When you close the door and see your reflection on the walls, floors and ceilings, "it's a totally sensual experience," she says. Stainless comes in lots of finishes—one that's been polished "multidirectionally," says Luce, is less likely to show fingerprints and other marks.

before

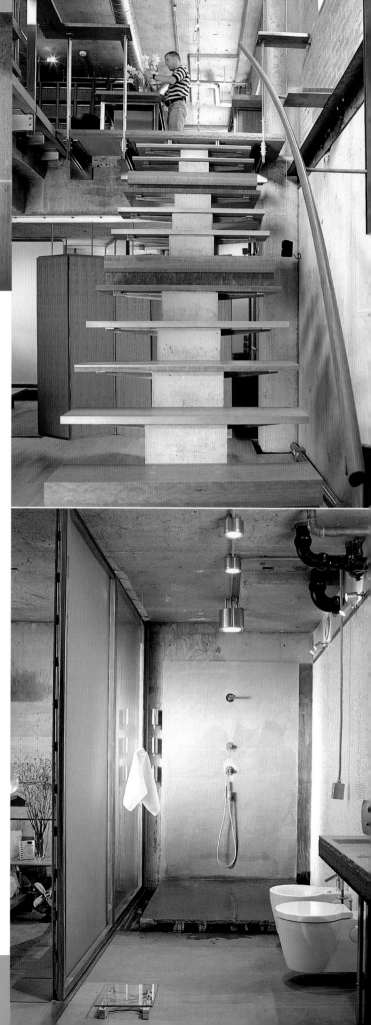

What the Pros Know About
Stone Finishes

It makes sense that a loft—a housing type developed in New York—would benefit from a supply of New York bluestone. Architect Jennifer Luce wanted the stone (a variety of granite) to have a matte finish—as a foil to the sleek surfaces around it. Unfortunately, "granite is generally polished before it gets to your supplier," says Luce. "There's this idea that polished stone is better." But Luce was determined to take the bluestone's sheen off. Luce ordered slabs in a variety of thicknesses (from two inches for kitchen counters to eight inches for the bathroom's pedestal sink). And she specified a variety of finishes. **For the kitchen counters, Luce chose a chiseled edge, "which gives the stone a truly natural look, as if it had just come out of the ground."** On the shower floor, made from a five-inch-thick slab, she used a flamed finish—"They literally hold a flame to the edge, to give it a rough, uneven texture," Luce explains. "It almost works as a pumice for your feet." For places "you're likely to touch with your hands"—including the shower wall and kitchen countertops—Luce chose a honed finish, smooth but still not polished. The bluestone, she says, "is dense enough that, even without the polish, it's not too porous." (A clear sealer, she adds, helps it resist stains). Thanks to the versatile bluestone, Luce says, "there's the same hue throughout the loft, but with very different effects."

SUBURBAN HOUSES

MYSTERY
HISTORY

WHAT THEY HAD AN OLD HOUSE IN A LEAFY SUBURB, WITH STRUC-
TURAL PROBLEMS AND A RUDIMENTARY KITCHEN, PLUS TWO
PORCHES THAT FELT LIKE TACKY ADD-ONS.
WHAT THEY WANTED A LUXURIOUS HOUSE THAT LOOKED "LIKE IT
HADN'T BEEN RENOVATED," WITH OUTDOOR LIVING SPACES AS
SUBSTANTIAL AS ITS NEWLY GRACIOUS KITCHEN.

HOW THEY GOT IT People walk in and say, "Is this the original kitchen?"
says Amy Wickersham, whose suburban New York house was built in 1890.
The question is understandable, since the warm colors and traditional
materials suggest a room that's been around for ages. In fact, there's
almost nothing left of the original kitchen. "It was much too small to eat
in," says designer Kevin Walz, "and was miles from the dining room and
further from the porch that would be used for summer dining, once we
screened it in." To make matters worse, the work space was a characterless
strip of counter on an exterior wall. "It made me understand why, in the
1950s and 1960s our mothers were clamoring for timesaving devices and
frozen foods," says Walz. "Who would want to cook facing the wall?"

From that dull kitchen, plus four surrounding rooms, the designer

before

created one big cooking/eating/relaxing space. But he broke it into zones—ghosts of the original rooms—that he describes as "emotionally and spiritually connected." Counters, islands and a double-sided, floor-to-ceiling china cupboard are the new walls. Cabinet fronts have pressed-tin panels; new wood was treated with milk paint, which creates a mottled (and antique-looking) surface.

"We created mystery history," says Walz of his approach. Throughout the house, the architect combined old and new in a way that keeps visitors guessing. In the living room, the floor, with exposed nailheads, is original to the house; the fireplace was brought over from Italy by a previous owner in the 1940s; and the French doors are brand-new. (Walz detailed the doors to look like bay windows, with solid panels of wood below knee height.)

The Wickershams worked in stages, which enabled them to live in the house throughout the three-year renovation. "We finished the outside before we did the inside, and when we were doing the kitchen, we kept the upstairs untouched," says Wickersham.

An artist, she says, "I don't plan a whole painting at once. It's an unfolding. The house was the same way."

Above: The dining table and chairs are "Anglo-Raj"—pieces created for Dutch and British colonials in India and Sri Lanka. Opposite: From her soapstone countertop and sinks, Amy Wickersham faces a cozy seating area—which makes her feel, she says, like she's in "a living room, not a kitchen."

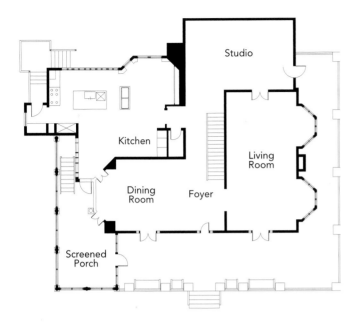

What the Pros Know About
Making Outdoor Spaces Feel Substantial

Kevin Walz wanted the Wickershams to be able to eat summer meals on a screened porch that was as comfortable as any other room. **To give the porch substance, he designed a floor of mahogany planks as wide as the kitchen's oak floorboards.** For supports, Walz used classical columns (instead of mere two-by-fours) to "give the room a presence." Fixtures are reminiscent of those used inside—the lantern could be in the dining room or foyer. The house also had a deep front porch of terra-cotta tile (it was "literally decaying," says Amy Wickersham) and a metal railing that was badly rusted. Walz replaced the railing with a wooden half-wall that makes the porch feel like a piece of the house's front facade brought forward, rather than a recent add-on. That was important to Wickersham, who says, "I wanted the house to look like it hadn't been renovated." The wall incorporates a series of wooden planter boxes, benches and stone-capped piers, all of which give it the heft of real architecture. The tiles were replaced with floorboards that complement the all-white house.

Above: Designer Walz wanted to give "dignity" to the master bathroom, so he replaced "very 1960s" tiles with squares of Carrara marble. The chubby mahogany shelf and vanity connect the bathroom to the bedroom, with cabinetry suggesting a ship's stateroom. Opposite: Walz made the lamp shades from buff-colored Corian. The bed is from Goa, in southwest India.

RENOVATION DREAM When Amy and John Wickersham hired designer Kevin Walz, he was living in New York City. A few months later, he won architecture's prestigious Rome Prize and moved to the Eternal City (where he has remained with his two daughters). Wickersham says she felt "simpatico" with Walz and never thought of replacing him, despite the physical distance between them. "If you're in sync mentally, it doesn't matter if you're thousands of miles away," she says.

RENOVATION CHALLENGE The first time she saw the two-story colonial, "it was literally falling apart—the worst house in the neighborhood," says Wickersham. The siding was badly rotted, and when that came off, it turned out the studs underneath needed replacing. "The problems around every corner were the worst-case scenario," she says. But, she adds, "renovation is like childbirth—when it's over, you forget how painful it was." Walz, who has never experienced childbirth, also looks on the bright side. He points out that removing the siding created the opportunity to insert proper insulation, "so the house is much better for having been in bad shape."

MIAMI TWICE

WHAT THEY HAD A REMNANT OF MIAMI BEACH'S PREWAR HEYDAY, WITH A WET BAR AS BIG AS A BEDROOM AND AN OUTDOOR DANCE FLOOR—BUT WITH STRUCTURAL PROBLEMS THAT WOULD HAVE CAUSED MOST OWNERS TO DECLARE THE HOUSE A "TEARDOWN."
WHAT THEY WANTED A HOME FOR ENTERTAINING, WITH THE ECCENTRICITIES OF THE ORIGINAL BUILDING—BUT AMENITIES LIKE AN INDUSTRIAL-STRENGTH KITCHEN AND A MASTER BATHROOM THAT PUTS POSH HOTELS TO SHAME.

HOW THEY GOT IT When Tom Healy and Fred Hochberg, who transformed Lillian Vernon, his mother's catalog business, into a public company, saw this sprawling dowager on one of Miami Beach's Sunset Islands, they were sure it was the weekend house they wanted. "Just like friendship, you know immediately," says Healy, an art dealer turned poet. But like some friendships, certain houses require a lot of work.

In the living room (above left), designer Alison Spear placed a pair of "very Miami Beach" Venini glass sconces around the fireplace and Iran Issa-Khan's shell photo on the mantel. The Parsons tables are mica. The house's former owners, portrayed as clowns (above right), bring a jovial touch to the kitchen, but the work area (opposite) is serious—with professional appliances and countertops of "graphite black" honed slate.

The house had been built by Vera Smith, a minor Coca-Cola heiress who died in the late 1990s. "When we bought the place," Healy remembers, "it was an abandoned wreck, overgrown and sinking." Evidence of Smith's eccentricities abounded: She had boarded up some of the doors and windows, to protect her from the sun's menace—"like a tropical Miss Havisham," says Healy. But like a time capsule, the house perfectly captured another era. The front yard was an outdoor ballroom, its dance floor paved in local Florida keystone (with a giant outdoor fireplace for chilly nights). Off the living room was a 80-square-foot wet bar.

Miami Beach–born architect Warren Ser, of Ser Design Associates, was hired to plan the house's recovery, which began, he says, as much with archaeology as with architecture—every time a wall was opened up or a ceiling came down, there were surprises (including a painted mural of the original owners dressed as clowns). But not all of Ser's discoveries were good. Structural concrete beams had spalled: The metal reinforcement rods—called rebar—in the concrete had rusted. "And when rebar rusts, it expands and can literally disintegrate the concrete," says Ser, adding that spalling is a common problem in humid south Florida. A 50-foot-long concrete beam, supporting the second floor bedrooms and bathrooms, had to

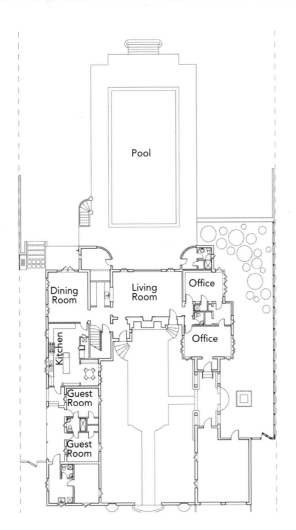

be restored (while temporary supports replaced it). The bars were removed and replaced, then reset into the beam with repair mortar. "Many clients," Ser says, "would have abandoned the project at this point." But not these clients. "The project was very much a team effort, with Warren as architect, Alison Spear as interior designer, Robert Parsley as landscape architect and yours truly as a very demanding, very opinionated, very involved amateur designer," says Healy.

With the major beam restored, Ser was able to take down walls, turning several small ground-floor rooms into what Spear calls "an industrial-strength kitchen," with new cabinets (some with perforated metal fronts) and countertops of honed Brazilian slate. Original Dade County pinewood floors were ebonized and covered in a tough marine finish. Ser oversaw refurbishment of more than 40 metal casement windows. "Everyone advised us, 'They're horrible; tear them out,'" says Ser. Once the metal had been restored, he replaced the old $\frac{1}{8}$-inch panes with $\frac{3}{8}$-inch hurricane-resistant glass.

RENOVATION DREAM The exterior wood trim had been painted many times over the years. "When we stripped the paint, we discovered that the original wood material was local cypress, naturally resistant to decay," says Ser. "We lovingly restored every inch of the wood and protected it with clear marine varnish to expose its original beauty."

Opposite: Restored casement windows reflect off a white-metal dining table from Miami's Luminaire. Chairs are by Antonio Citterio, and the chandelier (from Anthologie Quartet) unites stainless steel with crystal. The outdoor shower's slatted floor lifts up, revealing a stairway to the basement room that houses pool equipment.

before

What the Pros Know About
Freestanding Tubs

Tom Healy and Fred Hochberg wanted a glamorous master bathroom, with a freestanding tub as its centerpiece. Warren Ser combined several small rooms into the new master bath and placed an Agape "Spoon" tub at its center. But the tub, filled with water, and the new floor slabs of ¾-inch white Thassos marble were too heavy for the original 2-by-12 wood joists. So Ser added several steel I-beams to support the extra weight. His next problem was waterproofing. And not just under the tub. The entire room is a shower—with a "deluge" showerhead in the ceiling. "Code requires a waterproof pan under the shower area; here, we had to treat the whole bathroom as a shower," Ser explains. **Before setting the stone and glass mosaic tiles on the walls, he applied Laticrete 9235, a waterproof membrane system** (laticrete.com), which is applied as a liquid and sets into a continuous "skin." With the bathtub in the middle of the room, Ser had to run plumbing under the floor, mindful of code that requires drains to be no more than five feet from vent stacks that lead outside (usually via the roof).

Oh, must we dream our dreams
and have them, too?
Elizabeth Bishop

PICTURE WINDOW

WHAT THEY HAD A 1950S BUNGALOW THAT TURNED ITS BACK ON THE NATURAL WONDERS IN ITS OWN BACKYARD, AND WAS TOO SMALL AND LOW-CEILINGED FOR A COUPLE WITH TWO CHILDREN.

WHAT THEY WANTED A HOME FOR A BLENDED FAMILY, WITH BOTH A GREEN EXTERIOR AND A GREEN HEART, WHERE WINDOWS ARE ATTUNED TO LIGHT, BREEZES AND VIEWS.

In David Lake's living room, the eye is drawn outward (to the views) as well as upward: Overhead shelves are reachable only by ladder; outside, trees provide all the symmetry the street facade requires. Opposite: Panels of corrugated plastic form a cheerful clerestory, while creating less heat gain than glass.

Opposite: Lake took a
mix-and-match approach
to windows. Tilt/turn
models cant in at the top
for ventilation, and swing
open like doors for easy
cleaning. Other windows in
the house are "awning
style" or "double-hung."
Lake recommends Marvin's
Magnum double-hung
windows because one of
them costs only a little
more than two small
windows—and installing
just one window
simplifies the framing.

HOW THEY GOT IT Renovators who aren't architects may wish they could go from concept to design without an intermediary. But renovators who are architects have other problems. "When you design for yourself, there's nobody to tell you when to stop," says David Lake.

Lake, a principal of San Antonio's Lake/Flato Architects, rarely lacks for ideas. The firm has created some of the country's most innovative houses, finding gracious domesticity in the architectural vocabulary of farm and factory buildings.

For years, Lake lived in a 1,500-square-foot, 1950s bungalow in San Antonio. The house was cute but hardly an architectural showplace. Its front door opened right into the living room. Ceilings were a just-tolerable eight feet. The backyard was a wildlife refuge—woodpeckers, cardinals, owls and hawks nest there—but the house lacked a proper picture window.

But when Lake married Ellen Hicks, in 1996, their new family— including his son and her daughter—needed more space. (And they needed it quickly; the couple had borrowed a friend's house and didn't want to overstay their welcome.)

In less than a year, Lake transformed his bungalow into an asymmetrical cube of green stucco and corrugated metal—now 2,900 square feet, with ten-foot ceilings. Determined to take advantage of the backyard view, Lake could have created a wall of glass. But that approach would have been environmentally unsound, admitting too much heat in summer (and been a design cliché to boot). Instead, Lake thought about where he wanted to frame the view, where he wanted light and where he wanted breezes, and placed windows accordingly. The result is an asymmetrical composition—as Hicks puts it, "David isn't into pairs and squares." The closest thing to a glass wall is a row of floor-to-ceiling double-hung windows from Marvin. Called Magnums, they're the largest wood double-hungs the company produces. Says Lake, "You get the advantages of double-hung windows"—including cool air entering below and warm air exiting above—"but at a really nice scale." Other windows tilt open, and still others are fixed.

Inside, Lake gave in to Hicks's request for an oversize kitchen island. He was initially skeptical—"I call it the aircraft carrier," he says—but now he thinks it's perfect, "since everyone congregates in the kitchen anyway." With more than a dozen cats and dogs, the family has left no surface of the house unpunished. The Sheetrock wall along the stairway is especially prone to dents and scratches; after years of repainting it, Lake says, he wishes he had used a more durable material, like plywood. But he'll get his chance. Says Hicks, "When you're married to an architect, your house is always changing."

In the kitchen, a floor of recycled oak is wonderfully irregular: "There is rich color and pattern," observes Ellen Hicks. Rafters are exposed and a table base is made of pipes. Inexpensive stock cabinets were customized with perforated metal fronts.

What the Pros Know About
Six-Inch Studs

In a hot or a cold climate, insulation is the key to energy efficiency. David Lake had the exterior walls of his remodeled house framed with six-inch studs (two-by-sixes), instead of the traditional two-by-fours. **That provided extra space inside the walls for insulation.** The additional cost was minimal, since six-inch studs can be placed 24 inches apart (rather than the standard 18). That means you need fewer of them, Lake explains. The architect estimates that using six-inch studs cost him an extra $300. In addition, Lake says, two-by-sixes tend to be "a little straighter and truer than two-by-fours," making it easier to build walls that look good quickly.

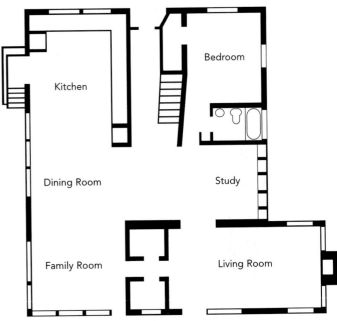

Kitchen

Bedroom

Dining Room

Study

Family Room

Living Room

Above: The upstairs bathroom is brand-new, but has a vintage look. Opposite: The stairs' pipe railing exemplifies what Lake calls the house's "quasi-techno-bungalow" demeanor. The architect got even more adventurous with the front door—which holds ephemera between sheets of acrylic.

RENOVATION DREAM After the renovation was completed, Lake added a large screened porch, which he says is now the family's preferred living and dining area (in San Antonio, he points out, it's possible to enjoy the porch ten months a year). Screened porches cost about half as much to build as enclosed, air-conditioned rooms, says Lake, so the project enabled him to enlarge his house significantly at minimal expense. A wide overhang provides shade in the summer months. "You don't go outside in the summer sun without a good hat," says Lake. His porch has a good hat.

RENOVATION CHALLENGE It seemed like a good idea to place a six-foot tall window next to the house's front door—at night, it served as a welcoming beacon, says Lake. But there was a cost in privacy: "You may be walking around in your robe when somebody drives by," he says. Since these photos were taken, Lake applied a frosted film to the glass, making the window less transparent.

BARN
AND
NOBLE

WHAT THEY HAD AN OLD BARN WITH DISINTEGRATING WALLS, A DANGEROUS BOWED ROOF, A CRAMPED KITCHEN—AND LIGHT, VIEWS AND BREEZES THAT THESE LONGTIME OWNERS WEREN'T WILLING TO GIVE UP.

WHAT THEY WANTED A SECURE SETTING FOR THEIR WORLD-CLASS COLLECTION OF CONTEMPORARY ART AND A KITCHEN SO GOOD-LOOKING, IT'S PART OF THE DISPLAY.

HOW THEY GOT IT A New York City couple had owned this house in a lush Philadelphia suburb for more than 20 years. Built as a barn in the 19th century and converted to a residence in 1937, it had thick stone walls and a ceramic tile roof. They thought it was in good condition but decided to treat themselves to a new master bathroom, a new powder room "and maybe a couple of closets," the wife recalls. A psychologist, she now admits that she was "in denial."

In fact, the stone walls were in such bad shape "that when you touched them, pieces fell off," she says. And the ceiling of the dining room was about to collapse. "Do you like that sway in the roof?" a friend inquired, not knowing that the sway wasn't intended.

Then, while the couple was in New York, a hose to the upstairs

washing machine came loose. By the time the leak was discovered three days later, much of the interior had been destroyed. And so began a gut renovation that included reorganizing much of the house. True, interior designer Carl D'Aquino and architect Francine Monaco (of Manhattan's D'Aquino Monaco) couldn't get the couple to change the upstairs layout. But downstairs they changed almost everything.

The kitchen, which was 6 by 20 feet, was functional but not roomy enough for socializing, says the wife. Other designers had proposed adding a new kitchen to the back of the house, but D'Aquino preferred to leave the old stone walls intact. Instead, he proposed turning a large (and rarely used) dining room into an eat-in kitchen. That meant the kitchen would be visible from the front door, an arrangement the wife wasn't sure she liked. But D'Aquino moved the appliances out of view and made the cabinets look like real furniture (ebonized to match the dining table). Besides, the eye isn't just drawn to the kitchen, but through it, to the park-like yard. He also moved a powder room to make room for a real foyer—large enough to display several important pieces of contemporary art. The biggest problem, the wife says, were the bills. "I thought my husband was checking them, and he thought I was checking them," she says, adding "But for perfection, it was worth it."

Above (from right):
The house is known to neighbors for the necklace in the yard—really a fiberglass sculpture by Karen Giusti; the new foyer contains a figure by Antony Gormley precisely where a tiny powder room once stood.

before

What the Pros Know About
Ebonizing Wood

When his client wanted to keep a dining set he didn't love, designer Carl D'Aquino suggested ebonizing it—covering it in a glossy black finish, which would make the pieces read as objects. "I get it every day," says Manhattan refinisher Paul Ebbitts of New York's Lacquerworks. "Clients wanting to keep pieces that designers don't love, so we ebonize them." **Ebonizing a dining set can cost $5,000 or more—it's very labor intensive, says Ebbitts.** The job begins with cleaning the pieces with alcohol, then sanding (use 220 sandpaper) to get the surface smooth. There's no need to sand all the way down to bare wood. ("It's like a car with lots of dings—they fill them in, and then they sand to get an even surface," Ebbitts says. "They don't have to expose the metal.") Then comes a layer of lacquer primer (Ebbitts recommends Mohawk's Clear Lacquer Sanding Sealer, which comes in spray cans). Finally, lacquer is applied in a spray booth. But if the furniture's old finish is badly "alligatored," Ebbitts says, more work may be required.

Dining Room

Pantry

Kitchen

Entry

Terrace

Living Room

Above: The bathroom has a marble countertop four layers thick and matching window surrounds. Opposite: The master bedroom wasn't moved, but it was updated with a teak bed and a matching custom-made TV cabinet.

RENOVATION DREAM Says the wife, "Knowing I was using a New York architect for a house in Pennsylvania, and knowing we wouldn't be there during the week, I asked friends who are very picky: 'Would you use your contractor again?' Out of 11 people, eight said no and three said yes. All three had the same contractor. That was it; we never put it out to bid."

RENOVATION CHALLENGE Surveying the second-floor layout, D'Aquino convinced the couple to leave their small corner bedroom for a larger room down the hall. The old bedroom would become a new master bath. Construction—which included knocking down a working fireplace in the old bedroom—took months, but when it was done, the wife changed her mind. "I couldn't leave the old bedroom. I had always slept there. I had always seen the pond and had that particular light and the cross ventilation." So everything had to be put back where it was, including the fireplace. Says the wife, "The mantel was never as pretty. And the new fireplace never worked as well as the old one."

SCENE STEELER

WHAT THEY HAD "A CRATE WITH VINYL WINDOWS," TOO MANY SMALL ROOMS AND NO ARCHITECTURAL POWER—IN A VERDANT SETTING WITH STUPENDOUS VIEWS.

WHAT THEY WANTED A LOFTLIKE SPACE WITH ARRESTING FEATURES—INCLUDING A DRAMATIC WALL OF WINDOWS WORTHY OF THE PACIFIC NORTHWEST VISTAS.

HOW THEY GOT IT Liz Dunn is a real estate developer; Bruno Lambert is a race-car driver—and they're both environmentalists. "We see people building these monstrous houses, with rooms for every possible activity," says Dunn. "They're creating uninhabited space, and they're also using up a lot of resources to do it," she says. At 2,800 square feet, the couple's own Seattle house seemed plenty big enough. Verdant surroundings, and views of Lake Washington and the Cascades, made it idyllic.

But architecturally, the house was no match for its setting. A vaguely modern 1970s box, the house, Dunn says, was "a crate with vinyl windows." Says architect Tom Kundig of the superlative Seattle firm Olson Sundberg Kundig Allen, "It was slapped together, the kind of place that starts to fall apart after ten years."

To make the most of their 2,800 square feet, the couple wanted a flowing, loftlike space. To connect the kitchen, living and dining areas, Kundig removed interior walls, including one that was load bearing. To take the weight of the second floor, he installed two steel columns. The columns are encased in maple cabinets (which bracket the new dining area); Kundig used push latches instead of handles so that the cabinets would recede. A "standard" wood stairway was replaced with a steel-and-concrete model of Kundig's design. Upstairs, a catwalk overlooking the double-height living room leads to an office and the couple's bedroom.

Above: A modest frame house is now a miracle in metal. Outside, new anodized-aluminum window frames match the black-painted cedar siding.

Maple cabinets (with steel reveals) surround two columns that replaced a bearing wall. The columns frame the dining area (with recycled-teak table and benches). Opposite: In the kitchen, concrete countertops are teamed with unconventional storage (including old file cabinets and lockers).

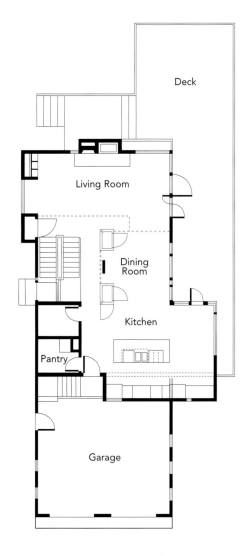

Deck

Living Room

Dining Room

Kitchen

Pantry

Garage

What the Pros Know About
Steel Stairways

"Steel is a heavy material," says architect Tom Kundig, "but its strength and rigidity allow for open, airy structures." **The house's old wooden stair would have looked clunky next to the new fireplace, windows and catwalk.** So Kundig designed a replacement, with wood handrails and concrete treads. Had the renovation involved removing the roof, preassembled flights could have been craned into position. Instead, the steel elements were welded together by a two-person crew. Then the treads and rails were added. ("The concrete treads dampen the resonance of the steel, while the wood is pleasing to the touch," says Kundig.) Steel can be polished or painted for a more finished look—but here Dunn and Lambert were happy to see weld marks, carbon residue and even the fabricator's writing. (Likewise, the few rust spots don't bother them.) Another plus: The metal stair will never squeak.

Outside, Kundig nearly doubled the size of the decks—"as a bonus," says Dunn, "we got rid of most of our lawn"—and replaced white-painted wood with cedar. New railings, of cedar with steel cables, match those on the interior stairway.

But the connection between indoors and outdoors, thanks to those off-the-shelf vinyl windows, was weak. So in the house's southeast corner, Kundig removed an old wood post and replaced it with a steel I-beam (which, Dunn says, was left primer-coat red "so it would really pop"). That meant he could make larger openings in the adjacent walls. Those openings now contain two huge doors—one hinged, one pivoting (at 5 feet by 14 feet, it weighs nearly a ton but moves easily on industrial hardware). Alongside the doors are a series of windows that salute the views.

Adding steel to a wood-framed house may sound like a mixed metaphor, but it is "no big deal," says Kundig, "You have an I-beam cut the same height as the wooden post, you weld steel plates to the top and bottom of the I-beam and you bolt it in. If the existing wood is solid, it's easy," Kundig says. In this case the steel beam allowed Kundig to make the house better, without making it bigger.

RENOVATION DREAM Most people these days want a room for every purpose. Dunn and Lambert took a different tack. "There's just two of us," says Dunn, "and we just don't believe we need a room for every different facet of our lives." Their entire first floor is now one flowing space. Says Dunn, "Bruno can be cooking, and I can be sitting at the dining table with my laptop, and the fire can be burning in the living room, and it's all connected. It's been an amazingly positive change in our lives."

HIGH BEAMS

WHAT THEY HAD A CLASSIC 1960S POST-AND-BEAM, ITS EXPOSED STRUCTURE SUGGESTING SOLIDITY AND STYLE—BUT WITH TOO MANY MATERIALS COMPETING FOR ATTENTION.
WHAT THEY WANTED A CLEAN-LINED HOUSE WHERE BEAMS ARE THE STARS THEY WERE ALWAYS MEANT TO BE.

HOW THEY GOT IT Lauren Golub and Bobby Turner had something in common: Both were raised in post-and-beam houses (she in Baltimore, he near Boston). So it's no surprise that, when they met and married, they were drawn to what they knew. Post-and-beam design, in which massive timbers (instead of the usual rows of two-by-fours) provide support, are preferred by modernists, who believe structure should be articulated, and anyone who likes a lot of light: When posts hold up the ceiling, walls needn't be more substantial than sheets of glass.

Opposite: **A procession of stucco-covered posts and exposed beams begins at the front gate, establishing architectural themes.**

The house the couple found, in L.A.'s Pacific Palisades, was a 1960s beauty. But the overall mood was cold, says Turner. The previous owner was an architect who specialized in large commercial interiors, "and that's the way he thought about the house." Tacky partitions resulted in what felt like "office cubicles," he adds. Too many materials—copper, granite and mirrored panels—competed for attention, recalls Lauren's sister, Tori Golub, a New York City interior designer who helped the couple come up with a strategy for rescuing the house. Says Lauren, "We decided to make it more streamlined and spacious."

The couple tore down walls they didn't need and added glass doors opening onto the newly landscaped yard. ("The gardens also

Ebony-stained floors paired with light rugs echo the dark-framed photographs on the fireplace (with new plaster surround by Tori Golub). Enhancing the house's open feeling, window frames seem to disappear into the floor.

required renovation," Turner says.) But the couple also added walls when doing so wouldn't impede light or views—in one case closing up a pass-through between the dining room and kitchen. One disadvantage of post-and-beam houses is that there may not be enough walls for hanging art, or even for furniture that can't easily "float." With the pass-through closed, the couple had a new wall, against which an Edward Wormley sideboard and a diptych painting nestle.

The couple wanted to replace the previous owner's carpet with hardwood floors, but the ceilings were already wood, and Lauren didn't want wood above and below—she was afraid the house would start to feel "like a log cabin in Montana." So on the ceiling between the over-

head beams, they installed Sheetrock painted the same white as the walls. (See box, page 166.) In the upstairs bedrooms, where the floors are carpeted, the wood ceiling remained.

It was Tori who suggested painting the metal brackets supporting the beams black, "because we liked the way they expressed how the house was put together," she says. She also designed the fireplace surround, which is made out of steel-troweled, tinted plaster. "It's the kind of material that, if you used it everywhere, would lose its specialness, but with dark wood floors and white walls, it's perfect." From his fireside reading area, Turner, a hedge fund manager, can see the refurbished yard and pool. "It's like a SoHo loft with the best outdoor terrace you can imagine."

Classic architecture calls for classy furniture, like a Chris Lehrecke daybed and a biomorphic coffee table by Paul Frankl.

before

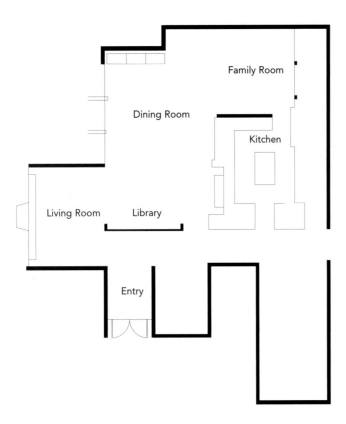

Family Room

Dining Room

Kitchen

Living Room   Library

Entry

What the Pros Know About
Ceiling Beams

"Beams express architectural character," says New York City designer Tori Golub. For that reason, homeowners often dream of exposing ceiling beams, and in many cases they can (although they may also expose wires, pipes and insulation in the process). **Once exposed, beams can be sanded, bleached, stained, painted, stenciled, roughened with tools for a hand-hewn look or, as in this house, simply left alone.** Here, the beams, chunks of lumber almost as hefty as railroad ties, were designed to be seen. (The walkway from the street has the same beamed structure as the interior rooms, previewing what's to come.) The couple was even happy with the "wishy-washy white stain" that the previous owner had applied. But there was a problem: The beams disappeared into an all-wood ceiling. When the couple covered the ceiling with Sheetrock, "the beams really popped," says Lauren Turner. Think of them as beaming.

Wide-open spaces are child-friendly, especially when there are curvy Arne Jacobsen chairs to climb on. Even to modernists, less isn't always more— adding a wall between the dining room and kitchen helped define both spaces. Opposite: An indoor seating area is flanked by new slate-covered patios with teak deck furniture.

APARTMENTS

CREATING VISUAL COMPLEXITY IN
A MANHATTAN HIGH-RISE

TURNING A VICTORIAN FLOOR-
THROUGH INTO A HOME FOR ART
IN SAN FRANCISCO

GIVING A PAIR OF NEW YORK
CITY PREWAR APARTMENTS A
LIKABLE LAYOUT

HIGH
STYLE

WHAT HE HAD A WHITE-BOX APARTMENT, WITH MANHATTAN VIEWS PROVIDING ALL THE ARCHITECTURAL DETAIL HE NEEDED, AND A PAINTER'S SENSIBILITY.

WHAT HE WANTED AN APARTMENT WHERE COLOR—AND TROMPE-L'OEIL EFFECTS—MAKE ROOMS SEEM BIGGER THAN THEY ARE, AND WHERE BOUNDARIES ARE DELIBERATELY BLURRY.

HOW HE GOT IT John Saladino is an interior design legend, and it isn't because he knows how to match fabrics to paint chips. Saladino is a master of making small rooms less confining, and big rooms more embracing, through illusion. When Saladino renovates, he changes perceptions of space as much as space itself.

Saladino's own apartment floats above midtown Manhattan. Some of the older buildings visible from his windows are so beautiful, he says, that his rooms didn't need additional architectural details.

But the layout of the two-bedroom apartment was too pat, and it was smaller than his last place (which had a 35-foot-high great room). Saladino used techniques he learned while studying painting at Yale to "de-physicalize the walls," he says, "to confuse your understanding of where each space begins and ends." In the living room, he had workers gouge out a ¼-inch-wide channel at the top of each wall. Then he painted the inside of the channel white (in contrast to the darker walls). The result: The ceiling seems to float over the walls, enlarging the room in the observer's imagination.

He uses curtains in the same way. In the living room, instead of simply covering the window, Saladino created dramatic, architecturally scaled drapes the full width of the room, and hung them nearly two feet from the glass. That creates a zone, behind the curtains but in front of

In John Saladino's living room, a limb supports a table (top left), and Central Park becomes a yard (top right). A folding screen (bottom left) softens a corner protrusion (and hides every designer's nemesis, the TV set). Curtains (bottom right) aren't window-size, but room-size. Sofas (bottom left) are underscaled, the designer says, to make the room seem bigger than it is.

A screen of metal mesh heightens the separation between living room and foyer (opposite top). A section of old copper roofing (opposite bottom), cut into the shape of a classical urn, gives a dresser a supporting role. In the guest bedroom (above), a small table behind the drapes hints at a room beyond.

the window, that is neither in nor out. Small pieces of furniture, strategically placed in that zone, suggest the presence of a room beyond. Likewise, to separate living room and foyer, he hung a metal mesh screen that extends about a foot beyond the wall on which it hangs— just enough, he says, to make you want to peer around it.

When it came time to furnish his apartment, Saladino chose seating in various sizes: "Some of your guests will be five-foot-two, and some will be six-foot-five," he says, "and everyone ought to be comfortable." He placed chairs facing each other, so guests can make eye contact—"It's the reason people linger at the breakfast table." But like a painter who composes background as carefully as foreground, he left a lot of the apartment empty. "The space," Saladino says, "is as important as anything in it."

Living Room

Guest Room

Master Bedroom

Foyer

Dressing

Entry

Dining Room

Kitchen

What the Pros Know About
Dark Paint Colors

John Saladino used dark paint colors for two different reasons: In one case, **his goal was to heighten the drama of entering the apartment.** So he painted the foyer (at rear in the top photo) "an elusive shade between midnight blue and aubergine." The living room beyond is a pale platinum. The result? "You enter a dark space and then, as you approach the living room, it seems gigantic, almost as if you've stepped outside." In the case of his library/guest room (opposite), Saladino wanted to create "a cocoon—the room you'd want to be in on a cold, dark night." For that, he chose a dark green paint. But when you do a dark room, he says, "you need to break it up with mirrors or a wall of prints or a huge painting." In this case, his huge painting is really a wallpaper panel glued in place. Because the sheets of paper are so much brighter than the wall, they appear to be free-floating.

MUSEUM
QUALITY

WHAT THEY HAD A FLOOR-THOUGH APARTMENT IN A VICTORIAN HOUSE, WITH DROPPED CEILINGS, A NARROW HALLWAY BETWEEN LIVING ROOM AND KITCHEN AND A LOT OF TACKY 1960S DETAILS. **WHAT THEY WANTED** A NEUTRAL BUT NOT BORING BACKDROP FOR A STELLAR COLLECTION OF ANCIENT, MODERN AND CONTEMPORARY ARTWORKS.

Previous spread: Hanging over Julie Dowling's fireplace is a 1907–08 Piet Mondrian landscape. The asymmetrical design of Dowling's limestone fireplace surround creates a symmetrical base for the painting. Opposite: The dining table (top left) is by Parisian designer Christophe Delcourt; the kitchen "chandelier" (bottom right) is from New York's Bone Simple. Even the bathrooms are galleries (middle right).

HOW THEY GOT IT An art dealer who sells important works from his home gallery, Steven Platzman may have a Cézanne hanging over his sofa and a priceless pre-Columbian vessel sitting on his desk. When Julie Dowling, an architect, began dating Platzman, she was intimidated by the museum-quality pieces. But when the couple decided to move in together, Dowling had a chance to become comfortable with the artworks, she says, "by designing an environment in which it felt like they belonged."

The couple's new apartment, a floor-through in a Victorian house in San Francisco, had been badly renovated in the 1960s. Dowling recalls its "lowered ceilings, textured walls, fake plasterwork and too-shiny oak floors." Still, she says, "this place was a big purchase for us. So we thought, Let's just paint the walls and move in. But as soon as we signed the contract, Steven started taking walls down. For two months, he forgot he was an art dealer, and he walked around with a sledgehammer. I would come home from work, and another wall would be gone. Luckily," she adds, "we had talked to a structural engineer, so we knew what we could take away."

The couple, now married, had decided that the apartment's front room—with windows that project out far enough to take in views of San Francisco Bay—would be their living room. But the kitchen was in the back of the house, at the end of a long hallway (a typical Victorian arrangement). Worse, the hallway passed between the bedrooms (on one side) and the bathrooms (on the other).

Dowling completely remade the hallway, adding shelves for

before

Breakfast Area

Kitchen

Master Bedroom

Guest Room

Media Room

Living Room

Dining

What the Pros Know About
Lighting Art

"Works on paper—prints, pastels and photography—require very low light levels," says Julie Dowling. "By contrast, metal and stone sculptures can be placed in full natural light, and their form is often highlighted by shadows." Many of the couple's works are paintings, for which they use a combination of incandescent and halogen lights—mixing floodlight and spotlight bulbs. **"The halogens are white, and they give you a very realistic depiction of the colors in the picture. Incandescents are slightly yellow, and can cast a work in a slightly more romantic light,"** says Dowling. But there are further permutations: "You may use a spotlight to focus light on a spot of red in a Corot, because red is very important in his work. Floodlights provide a more general light. Some paintings, because of the level of varnish on the surface, or simply because of the glass, require a floodlight, rather than a spotlight, so you don't get a pool of reflection on the painting." Many of the lights in the couple's apartment are on tracks, permitting adjustment of not only angles but positions. "People are scared of tracks, but if they're thin enough and very high quality, they can give a sophisticated, clean look to a ceiling," Dowling says. Not surprisingly, all of the lights are on separate dimmers. "You may want to bring up the halogen floods and lower the incandescent spots," the architect explains.

FRIDA KAHLO

Platzman's collection of art books, and through-the-wall niches. The niches, faced in Plexiglas, contain pre-Columbian vessels. (Unable to crash to the floor, the pieces would survive a small earthquake, Platzman predicts.)

With the hallway more of a space to linger in, the living room and kitchen now feel linked, rather than separated. It also helped that Dowling put as many of the mechanical systems as she could under the apartment's floors (now covered in dark walnut). Burying pipes and wires let her raise the ceilings. The hallway, which had been nine feet high, is now closer to 12 feet, and so, proportionally, it doesn't feel as long, the architect observes.

Dowling (who worked for Michael Graves before opening Dowling Kimm Studios in San Francisco) simplified the apartment's lines. The fireplace surround, which had been a shiny marble, is now a flat slab of Italian limestone, a material Dowling likes because of "its evenness in color and texture." But she didn't eliminate signs of age. "I didn't want an interior that felt foreign in a 19th-century Queen Anne Victorian structure," she says. Having raised the ceilings, Dowling replaced six-foot-eight-inch doors with eight-foot models. With those new proportions, Dowling says, the place "is more, not less Victorian" than before the renovation.

One typical Victorian element—brightly colored and patterned surfaces—is absent. Dowling needed to create a neutral backdrop for art, which she did by painting the walls the color of rice paper and installing a track lighting system (see box, page 181). The art in the apartment changes all the time, which poses a challenge for Dowling. Her strategy? "I may change the pillows, because I'm highlighting a particular intensity of color, or I may bring in a new group of books that reflect the cultural meaning of a particular work. The goal is to treat each room as a work-in-progress."

SOFT TOUCH

WHAT THEY HAD TWO APARTMENTS COMBINED INTO ONE WITH A RESULTING AWKWARD LAYOUT BUT WITH STRONG PERIOD DETAILS AND GLORIOUS EAST-FACING WINDOWS.

WHAT THEY WANTED A FAMILY HOME THAT WOULD MAKE THE MOST OF OLD MOLDINGS AND MORNING LIGHT, AND WOULD BE CALMING TO COME HOME TO.

Opposite: Marcel Wanders's string chair blurs the border between hard and soft. Throughout the apartment, simple sconces (of fabric over wire frames) designed by Calvin Tsao and Zack McKown provide most of the lighting. Above: The dining table, also custom-designed, conceals electric outlets and telephone jacks, allowing it to double as a work space.

HOW THEY GOT IT New York City architects Calvin Tsao and Zack McKown know a good thing when they see it. And this apartment, in a prewar building overlooking Central Park, had plenty of good things—period details worth preserving. The goal of their renovation, says Tsao, was to "honor the apartment's characteristics, not clip its wings."

Still, those wings needed some preening. Functional problems—including a lack of closet space and a nearly unworkable kitchen—abounded. And the brightly colored decor wasn't restful. Says Tsao, "The family wanted something calming to come home to."

A dancer before he became a designer, Tsao says he is attuned to how people move through spaces. That helped him finesse the floor plan. Tsao and McKown rounded a corner of the foyer, making it easier to get to and from the kitchen. And they enlarged the opening between the living and dining rooms—enough to help the rooms flow together, while still allowing each to "hold its own." To further improve the flow, Tsao and McKown custom-designed most of the furniture. Not because they were breaking stylistic ground, Tsao admits, but because they wanted to be able to fine-tune proportions: The pieces are substantial yet compact enough to keep the rooms from feeling cluttered.

The designers lavished just as much attention on the private spaces, including a master bedroom, study ("It's the place for a tête-à-tête," says Tsao) and bathroom that together total only 600 square feet. The owners, a couple with a young child, wanted a significant amount of storage. But building new closets would have meant shrinking the already-compact rooms. Instead, Tsao and McKown built tall cabinets that read as large pieces of furniture, taking up floor space but letting the rooms retain their full dimensions. The bedroom is ethereal—"Its colors are all sky colors," Tsao explains. As a result, "with light pouring in, the architecture almost disappears."

In the public spaces, Tsao used ten different paint colors, keeping the rooms pale while heightening the effect of sun streaming in through the windows. (Perpendicular walls are painted different colors, suggesting sunlight hitting them at different angles.) The darkest of the ten colors, an apple green, was reserved for the restored moldings, effectively framing the couple's art. That way, the period molding became part of Tsao and McKown's contemporary composition.

What the Pros Know About
Banquette Seating

A banquette, to architect Zack McKown, is "a piece of furniture that has architectural significance—it helps to shape the space." **It can also make a kitchen at once more glamorous and more convenient.** Convenient, because with a banquette there's always room for extra guests at dinner. (Children love squeezing into them.) Glamorous, because they recall posh restaurants. Which is why the easiest way to have a banquette made is to visit a restaurant-supply company in your city. (In New York City, the Bowery has dozens of such firms.) **For a more custom look, have a carpenter build a bench or a box, and an upholsterer make cushions.** Many banquettes have storage underneath; the one in this apartment stands on legs, which makes the compact room feel a little more spacious.

In the master bedroom, the floating TV (and the rosebuds in the chair by Shiro Kuramata) are apparitions. Tsao and McKown designed the night tables with lips, so bottles and glasses don't slide off during the night.

SMALL PLACES

BRINGING ZEN SIMPLICITY
TO A CONFUSING NEW YORK
CITY SPACE

TURNING A WING OF A LOS
ANGELES MANSION INTO A
HOME OF ITS OWN

LATCHING ON TO A DECREPIT
PENNSYLVANIA FARMHOUSE

MAKING A TINY MANHATTAN
APARTMENT THINK IT'S A LOFT

LOFT
HORIZONS

WHAT SHE HAD A SMALL, ODDLY SHAPED APARTMENT IN A FORMER FACTORY BUILDING, WITH A DROPPED CEILING SO LOW AND SO COMPLEX THAT BEING IN THE PLACE FELT DISCONCERTING.

WHAT SHE WANTED A LIGHT-FILLED, PEACEFUL AND CONTEMPLATIVE SPACE WHERE WALLS, FLOORS—AND EVEN THE CEILING—WORK IN CONCERT.

HOW SHE GOT IT The ceiling is the last part of a room some people notice. But when Linda O'Keeffe, *Metropolitan Home*'s director of design and architecture, was considering buying a 680-square-foot apartment in New York's Greenwich Village, the ceiling was one of her chief concerns. The unit is on the top floor of a 19th-century factory building, where it should have been airy and light-filled. But the previous owner had dropped the ceiling, to as low as seven feet in some places. What's more, the ceiling height changed so erratically that "it made the whole space feel chaotic," says O'Keeffe.

Opposite: An African feather headdress—of the type worn by tribal chiefs in Cameroon—hangs over the fireplace. Linda O'Keeffe stores firewood in a recessed stainless-steel box.

195

before

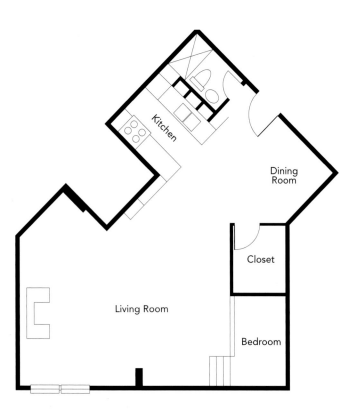

Kitchen

Dining
Room

Closet

Living Room

Bedroom

What the Pros Know About
Baseboards

In traditional construction, baseboards serve a simple purpose—concealing imperfections where walls and floors meet. But modernist architects insist on precision (and eschew concealment). So many go to the other extreme, replacing baseboards with a channel (sometimes called a reveal) in the wall immediately above the floor. But it's a difficult detail to achieve in new buildings, and almost impossible in old ones. **Linda O'Keeffe found a middle ground: a way to create baseboards that weren't reveals but weren't protrusions, either. She bought inexpensive sheets of metallic Formica, at four by eight feet, and had them cut into ten-inch-wide strips.** The strips were simply glued onto the wall, using epoxy, giving her the look (without the bulkiness) of baseboards. One wrinkle: If your wood floor has been sanded down many times, the old baseboard may sit as much as half an inch above the surface of the floor. In that case, the edge of the old floor may have to be gouged out before a new baseboard can be installed at floor level.

"Ellen's Brackets," by designer Ali Tayar, support shelves of ³/4-inch-thick Lucite. Be sure to get the edges of the Lucite polished, O'Keeffe advises. An Ethiopian basket rests on an Eero Saarinen table. Opposite: Folding closet doors were eliminated. The wall enclosing O'Keeffe's new closet (to the left of her "train compartment" bed) continues straight up into the well of a skylight. Curtains are from the late great Mary Bright.

O'Keeffe figured that the previous owners must have dropped the ceiling to hide wires, pipes or ducts. So it was a big relief when her "demolition guy" climbed onto an overturned bucket and punched a hole in the false ceiling. There was nothing up there except crisscrossing beams. That's when O'Keeffe knew the place had possibilities.

But the apartment's odd, angular layout wasn't easily resolved. O'Keeffe called Chip Bohl, a Maryland architect, for help. "I didn't think I had a lot of options with a space this small," she said, "but Chip showed me I did." They reduced the number of interior walls (only the bathroom and a walk-in closet are enclosed, and those are entered through doors of translucent glass). The "bedroom" is a kind of "train compartment," reached by a pull-down ladder. (Under-bed storage is accessed from the walk-in closet, so the front of the platform is solid.) O'Keeffe's kitchen cabinets, of ebony-stained oak, make 90-degree turns into the living room. Says Bohl, "That blurs the line between where the kitchen starts and where it stops—expanding the kitchen and the living room at the same time."

Above: What had been a gallery kitchen now extends into the living room, thanks to cabinets that turn corners. Opposite: Stainless-steel shelves were set into an unexpectedly thick wall. A glass panel in the corner brings light to the bathroom.

Above (from left): Boxing in beams, architect Chip Bohl created overhead sculpture. David Weeks's lamps illuminate O'Keeffe's raised bed, with its pull-up ladder. Opposite: Her Roman tub is covered in glass tiles, which O'Keeffe chose for their handmade look. Disliking shower curtains, she opted for a sheet of tempered glass.

Because she was her own general contractor, and on-site every day during the demolition phase, O'Keeffe was able to take advantage of opportunities that she wouldn't otherwise have been aware of. When she discovered that the wall between her kitchen and bathroom was 15 inches deep, she had stainless-steel kitchen shelving recessed between the studs. The original fireplace, though more than a foot from the outer wall of the apartment, had been connected to that wall with Sheetrock. O'Keeffe eliminated the superfluous material, re-creating the fireplace to as a freestanding sculpture.

Bohl worked to make O'Keeffe's ceiling less busy. He aligned a new living room wall with the well of an existing skylight. That way, when the light comes through the skylight, "there's no offset and no shadow," he says. The move "also had the effect of increasing the height of that wall, literally, and also creating the illusion that the wall continues on beyond the ceiling." Elsewhere, Bohl removed the old dropped ceiling, and treated the overhead beams as objects passing through the space. "We let the beams float, rather than lower the ceiling to try to conceal them," he says. The result: more of the lightness and airiness that O'Keeffe thrives on.

SMALL MIRACLE

WHAT HE HAD A 750-SQUARE-FOOT CHUNK OF AN OLD HOUSE, WHERE SMALL ROOMS WOULD HAVE TO DO DOUBLE DUTY, AND A BACKYARD TOO PRECIOUS TO BUILD ON.

WHAT HE WANTED A HOME WHERE ROOMS SEEM BIGGER THAN THEY ARE, WHERE MODERNIST INVENTIONS COEXIST WITH PREWAR DETAILS, AND WHERE AN OUTDOOR LIVING ROOM IS AS COMFORTABLE AS THE ONE INSIDE.

HOW HE GOT IT Steven Shortridge's house in Venice, California, may be tiny, but it thinks big: It was once part of a much larger house, in Hollywood. About 70 years ago, when that house was demolished, someone—Shortridge doesn't know who—salvaged a 750-square-foot section and moved it to a lot in Venice. The spin-off consisted of the parlor and dining areas of the original building. Its walls, in keeping with the scale of a much larger building, are almost 14 inches thick.

Shortridge loved the quaint castoff but needed to reorganize the spaces to make it livable. And he chose not to add on, because, he says, the house and the lot were in just the right proportion. (There was space for a terrace he calls "an outdoor living room," with a fireplace and

Above: **Steven Shortridge** clad the fence outside his house in nearly indestructible concrete panels. Opposite: The former dining room's built-in breakfront is now living room storage.

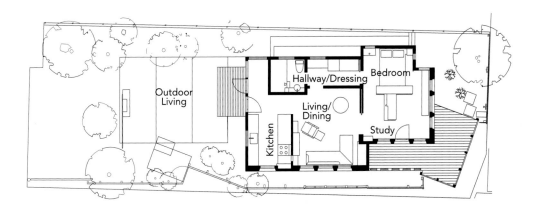

Floor plan showing: Outdoor Living, Kitchen, Living/Dining, Hallway/Dressing, Bedroom, Study

custom-made mahogany seating.) Inside, Shortridge looked for ways to make spaces seem bigger than they are. In one case, where he closed up a doorway, he kept a vestige of it—a slit about 5 inches wide and 30 inches high—that provides a glimpse of the space beyond. A wall made out of storage cubes open on both sides separates study and bedroom areas (carved out of the old parlor), allowing each room to "borrow" the dimensions of the other.

The old house's dining room became Shortridge's indoor living room. He tore off the old ceiling, which gave him an extra foot of headroom. He installed a new ceiling of Douglas fir tongue-and-groove boards, similar to floorboards. (Like floorboards, the ceiling planks were "blind nailed"—meaning the nails go in from the side and become invisible.) When he removed the old ceiling, Shortridge discovered that the outside wall thinned at the point where the ceiling had rested on it. With the ceiling gone, there was a kind of ledge, where he installed halogen lighting. The bulbs create an upward glow, making the new ceiling seem even higher than it is.

In the reorganized house, each of the spaces does double duty— there's the study/bedroom, the living/dining area and the hallway/ dressing area. In the kitchen, Shortridge created cabinets out of inexpensive medium-density fiberboard (MDF). Some of the surfaces were covered in wood veneer, but most others were left exposed—a contrast of smooth and rough that suggests, Shortridge says, that a finished material was cut away. The clean-lined cabinets coexist with the house's original built-ins, which Shortridge restored, part of a conscious mixing of old and new. Vintage baseboards were retained, but near the ceiling Shortridge created a modernist detail: a small reveal that suggests the walls and ceiling slide past each other without touching.

Clockwise from top: Storage cubes separate bedroom from study— without allowing either to feel claustrophobic; kitchen cabinets are made of medium-density fiberboard; "Because it's consistent all the way through, MDF can be used sculpturally," says Shortridge; in the dining area, what could be a modernist cat door is just an architect's way of saying "there's a world beyond this room."

before

What the Pros Know About
Stucco

These days, it's common to cover houses in artificial stucco, a sprayed-on, synthetic material that can cause serious problems. (There is no way for water that gets trapped behind the impervious material to seep out, so walls have been known to rot away from the inside.) **Steven Shortridge's house was built of real stucco— Portland cement, sand and lime into which powdered pigment was mixed into the stucco.** To achieve what he calls "New Mexico dirt orange," Shortridge needed a lot of pigment—and if you add too much, you can "overwhelm the cement base and break down its ability to dry properly," he says. For a very intense color, "hire a good person." Even so, mixing the stucco "took a lot of trial and error," Shortridge says. Often, stucco is applied with a float device, which creates a rough surface; Shortridge wanted his stucco smooth, so he had it applied with a steel trowel. The effect, he says, is almost like Venetian plaster. And yet there's a resemblance to adobe. To make his house look more substantial, Shortridge removed a layer of decorative tiles atop the walls; they made the surface seem thin, he explains. This way, the stucco suggests a substantial volume. Indeed, Shortridge says, stucco, because it's an earthy material, looks best when it's used on surfaces that are (or appear to be) connected to the ground.

NEW LEASE

WHAT SHE HAD A TINY FARMHOUSE, LITERALLY FALLING APART, IN THE MIDDLE OF A 500-ACRE NATURE PRESERVE AND AN IDEA FOR SAVING IT FROM DEMOLITION.
WHAT SHE WANTED A ROMANTIC RETREAT IN THE WOODS (JUST MINUTES FROM THE CENTER OF PHILADELPHIA), AND NEW ROOMS THAT DON'T UPSTAGE THE CHARMING OLD ONES.

HOW SHE GOT IT Sitting in architect Martha Finney's 200-year-old farmhouse, sun pouring through windows cut into 16-inch-thick stone, it's hard to believe that the building's owner wanted to tear it down. But the house had long stood empty while vandals carted off everything of value. The first time Finney saw it, it lacked plumbing or electricity and was filled, she says, with piles of trash.

The house sits in the middle of an environmental education center on 500 acres in Philadelphia; the center's trustees saw it as a nuisance. But before going ahead with the demolition, they decided to offer an 80-year lease (at a nominal rent) to anyone who could figure out a way to make the building habitable. Finney, a painter, entered the competition, which she won by proposing to keep the old house almost exactly as she found it. Given its "overpoweringly creepy condition," she says, "my friends thought I was crazy."

But Finney had a plan. "From the street," she says, "I didn't want the house to look like it had been hit by an architect." Instead, she designed a 12-by-28-foot addition, with a kitchen on the first level and a bathroom on the second. The new wing, which is mostly invisible from the street, would form a kind of service module. That meant she didn't need to bring new systems into the creaky old building. Aesthetically, she

before

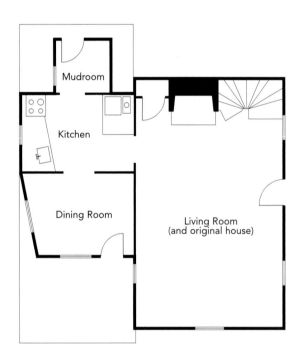

Mudroom

Kitchen

Dining Room

Living Room
(and original house)

"I knew I wanted a bright kitchen," says architect Martha Finney, who chose aqua (for concrete floors) and lime (for the room's clapboard walls). In plan, the new kitchen and dining area "cling" to the 200-year-old building.

What the Pros Know About
Staining Concrete

Martha Finney figured she could get away with inexpensive concrete floors in the new part of her house—after all, she could stain them bright colors (consistent with her painter's sensibility). **For that, she chose Kemiko, a concrete stain designed to create an almost marblelike finish.** Containing hydrochloric acid, the chemical seeps into the surface and is usually protected by an acrylic "wax." Finney selected aqua. In the summer, with leaves on the trees, "the room literally melts into the outdoors," she says (kemiko.com).

figured, "the old and the new would bring out each other's strengths."

In the old house, she was more archaeologist than architect. Using a sledgehammer, a task she found "deeply satisfying," Finney discovered the original fireplace behind three layers of brick. And the white pine floors were well preserved beneath several strata of linoleum and carpet. New window frames were inserted into the thick walls, and a tiny corner stairway was rebuilt.

The new wing is as modest as its partner. Its chemistry-lab counter-tops and walls of clapboard siding make it seem like it could have been built any time in the last 100 years. A radiant heating system beneath the floor means there are no vents or baseboards to give away the building's vintage. A standing-seam metal roof unites the old and new buildings. However, each retains its own identity. The new wing is so self-contained, "it looks like it landed, and could take off again at any time," jokes Finney. As for the old wing, she adds, "From the street, people still think the house is abandoned."

RENOVATION DREAM Finney couldn't afford to replace the wooden floors of the old house, nor did she want to sacrifice the 200-year-old boards. Instead, she coated them in a light gray paint. "The paint reflects light on even the dreariest day and turns a motley floor into a soothing patchwork of texture. I recommend it to anyone who will listen."

Above: A window in Finney's living room is also an exit to the porch. The bedroom and bathroom are upstairs. Opposite: An oval hospital-style curtain and ceiling-mounted showerhead update a vintage tub. The tub, too big for Finney's winding stairs, was lifted through the window.

What the Pros Know About
Old and New Siding

Martha Finney's old house was built of Wissahickon schist, a local stone, covered with a rough mortar of cement, horsehair and lime. Finney had the walls repaired but otherwise didn't move a single stone (or hair). For the exterior of her addition, she wanted something with just as much character, which is why she chose wood—cedar instead of the less expensive (and less durable) pine. **Finney had the wood stained, not painted, because stain won't peel or chip as it ages.** (Still, it's not immortal; every eight to ten years the walls need to be washed and restained, Finney advises.) She opted to have her boards stained at the mill. Prestaining—dipping the boards in vats—seals them on all sides, something postconstruction staining doesn't do. It also means there's less work at the site. Once the boards are up, all that's left to do is to paint the nailheads, though Finney, in pursuit of picturesque imperfection, decided to leave hers bare.

Finney describes the effect of the house as "hazy familiarity—it reminds you of something you've seen before, but in a slightly altered state."

PERFECT
FIT

What the Pros Know About
Fin Color Ply

All plywood consists of layers of wood glued together. Normally, only the exterior layers are veneer quality; the interior layers are rough. That means that when you cut the plywood, exposing the interior layers, the edge looks "crummy," James Gauer says—and if you're trying to make nice furniture (like the dining table and flanking credenzas in his living room), you'll need some kind of cap. **In the case of Fin Color Ply, all the layers are veneer quality, so you get a perfect-looking edge (requiring, at most, sanding and waxing).** The product comes with its top and bottom layers either natural or stained in a variety of colors (Gauer chose brown). It is about three times as expensive as regular plywood, but the finished edges mean "you save a lot on labor," Gauer says. And if you plan carefully (as Gauer did), you can do a lot with a four-by-eight-foot sheet (fincolorply.com/clrply).

WHAT HE HAD A TINY APARTMENT WITH GREAT VIEWS THROUGH CLASSIC CASEMENT WINDOWS, AND THE KIND OF GEOMETRY (A DOUBLE SQUARE) THAT ARCHITECTS FIND IRRESISTIBLE.

WHAT HE WANTED THE SAME TINY APARTMENT, BUT WITH THE OPEN FEELING OF A LOFT, A BIGGER KITCHEN AND A BEDROOM BUILT OF CABINETRY RATHER THAN WALLS (BECAUSE WALLS OCCUPY PRECIOUS INCHES).

HOW HE GOT IT At 500 square feet, architect James Gauer's apartment is about the size of a two-car garage. It was built as a one-room studio with a Murphy bed in a closet. The kitchen was precisely three feet long. But not only did the apartment have views of several Manhattan landmarks, it was a double square—which, Gauer says, is "the kind of geometry architects dream of." As a double square, the unit would divide neatly in half.

Gauer (of Gauer & Marron Studio, with offices in New York and British Columbia) knew he had to have one generous space, or else the apartment would feel like a warren. And so he set aside one half of the apartment for the living room. Fitting everything else into the other half was the real challenge.

First he created a small entry hall, which, he says, is "extravagant for an apartment this tiny." Then he ripped out a closet, which (although it wasn't large) allowed him to triple the length of his kitchen counter.

That left room for a bedroom not much bigger than the queen-size bed inside it. Cleverly, Gauer built the enclosure not from Sheetrock, but from cabinets—a series of compartments to replace the closet space he'd lost. "There simply weren't adequate inches to build a Sheetrock partition and then add closets or cabinetry," he explains. Given his every-inch-matters approach, he had the cabinets custom made, but "in a more forgiving space, something similar might be achieved just by ganging

Terrace

Living Room/
Dining Room

Entry

Bath

Bedroom

Kitchen

some stock cabinets together," the architect suggests. Gauer also gave the bedroom transom windows, one opening onto the living room, the other onto the foyer, to keep the room from feeling claustrophobic. Viewed from outside, the room looks like a little building—which (miraculously) makes the small apartment feel more like a spacious loft.

Even in the 12-foot-wide living room, Gauer had to choose furniture carefully. He ordered sofas just 28 inches deep (as opposed to the more typical 34). He made desks, tables and storage pieces out of Fin Color Ply (see box, page 218), which allowed him to customize their dimensions. Still, there were problems.

RENOVATION CHALLENGE The sofas Gauer designed for his living room are only 28 inches deep, which, he says, "means they should have glided perfectly through the 30-inch-wide door to the apartment. But for reasons I will never understand, the mover turned the first one on its side and slid it into the door frame, where it promptly became immovable. It ultimately took eight people, crammed into a tiny space, to remove not only the entry door but also the adjacent closet doors from their hinges before the sofa would budge. Luckily, nothing was seriously damaged."

RENOVATION DREAM "I had wanted a smart little stainless-steel kitchen ever since I saw one in a photo of Tom Ford's Paris apartment. But I consigned this to the realm of impossibly expensive dreams until a friend turned me on to a company in Chinatown that makes stainless-steel restaurant kitchens. 'It's not top-of-the-line quality,' he warned me, 'and you have to keep the drawings simple, but they're really nice people, and the price is right.' So I contacted Peter Chang at Bowery Restaurant Supply, sent him drawings, one for upper cabinets and one for lower, and he came back to me with a price of just over $3,000—including a counter with an integral sink. This was a serious bargain, fabricated on time and with no hidden costs."

What the Pros Know About
Steel Casement Windows

James Gauer's 1920s building, in New York's Tudor
City, was constructed with steel casement windows,
their familiar grids recalling an earlier era. The old
windows were made of thin pieces of steel designed
for single panes (which were held in place with putty).
Several companies, including Hopes (hopeswindows.
com), make casement windows. Gauer turned to New
York's A&S Window Associates (aswindowassociates.
com), which occasionally accepts small orders. Many
old casement windows pivot outward; to make them
safe for small children, new ones can be designed to
tilt open on top. Gauer didn't need insulated glass—
his windows are indoors. **But most new casements
are designed for energy-efficient double panes; the
downside is that, to hold the extra layers, the
frames need to be thicker than in vintage models**
(reducing the ratio of glass to metal). Gauer ordered
his windows unpainted and regrets it—A&S offers factory paint jobs in virtually any color. The windows are
delivered without glass; a glazier may charge as much
as $40 to cut and install each pane.

MASTER
WORKS

HELPING PHILIP JOHNSON
REALIZE A 50-YEAR-OLD VISION
IN CONNECTICUT

CELEBRATING CASTOFFS
ON A WASHINGTON STATE
INDIAN RESERVATION

GIVING FRANK LLOYD WRIGHT A
ROSY GLOW IN ARIZONA

HONORING

PHILIP JOHNSON

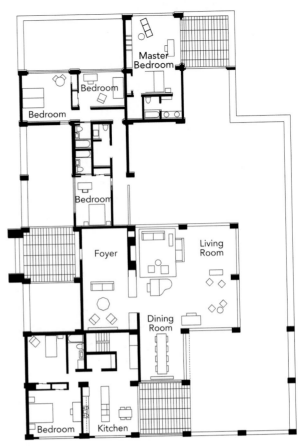

WHAT THEY HAD A PHILIP JOHNSON HOUSE DESIGNED IN THE 1950S BUT OVERLOOKED AND OVERDECORATED EVER SINCE— AND THE ADVICE OF THE NONAGENARIAN JOHNSON HIMSELF. WHAT THEY WANTED A PLACE THAT WOULD USE MODERN TECHNOLOGY (INCLUDING INSULATED WINDOWS) TO HELP THEM ACHIEVE JOHNSON'S VISION.

Above: A chaise by Ludwig Mies van der Rohe anchors a corner of the living room. Mies's influence is seen in the house's simple geometry, but its brick piers (suggesting columns without capitals) are classic Johnson.

HOW THEY GOT IT One thing Pamela Valentine learned about her husband, Bill Matassoni, during the renovation of their Connecticut house, is that "he's really a repressed architect." What renovator isn't? Still, when Matassoni, a management consultant, and Valentine, a matrimonial lawyer, began renovating this 1956 house, he couldn't let his architectural fantasies run wild. That's because the house already had an architect—Philip Johnson. It was Johnson's vision that guided the renovation.

The house, featured in dozens of architecture books (almost always in classic black-and-white photos), is a series of glass-and-brick pavilions. Even the floors are brick, which gives the architecture a solidity absent from many 1950s buildings.

But until Matassoni and Valentine came along, Johnson (now in his late 90s) was in better shape than the house, which had declined both physically and aesthetically. When the couple first saw it, it had already languished on the market for more than three years.

The original owner, Eric Boissonnas, played a pipe organ, and so the main room is 32 by 32 feet and 18 feet high. Friends of Matassoni and

Throughout the house, the couple kept furniture low and simple to avoid competing with the architecture. The "formal" dining room features "Brno" chairs, by Mies van der Rohe, and a Persian mosaic that belonged to the house's previous owner, renowned interior designer Jay Spectre. Opposite: An overhead trellis softens the transition from garden to front entrance.

What the Pros Know About
Interior Brick

In the Matassoni/Valentine house, most surfaces that aren't glass are brick. Brick not only forms the interior walls, but also the floors in nearly every room except the kitchen. (Wisely avoiding too much repetition, architect Philip Johnson chose different patterns for the walls and floors.) When the couple bought the house, the floors "had been lacquered and were very shiny. Awful," Matassoni recalls. **After sanding down the floors, workers used razor blades to remove the remaining lacquer.** The job was so time-consuming, Valentine says, that "they were finishing it when the moving van arrived." But Valentine loves having brick underfoot: "It's softer than people think, and it retains heat—it's warm in winter." As for the brick walls, she says: "They turn white anytime they get wet, so if you pay attention, they'll always tell you if there's a problem."

Above: A glassed-in hallway leads to the couple's bedroom. Opposite: The kitchen has a new slate floor, but the metal cabinets and pulls are almost 50 years old. (The stainless-steel lip below the upper cabinets conceals lighting.)

Valentine's call it the "ohmigod room." But overdecorating had made it the "ohmigosh room." Ditto for most of the house.

The couple kept the elements that worked—including the original St. Charles kitchen cabinets. But, Matassoni says, "we fearlessly tore out the glitzy bathroom. No matter how valuable the marble," he says, "it simply didn't belong in Johnson's house."

The couple devoted much of the renovation budget to such mundane matters as making the brick floors less glossy (see box, page 231) and replacing the single-pane 1950s windows with double-panes. With the old single-panes, previous owners had needed to use curtains for insulation in both hot and cold weather. But limits on the size of double-pane windows meant the couple had to add occasional vertical mullions. Says Matassoni, "Johnson came over and I asked him what he thought, and he immediately said, 'Do it.' He knew his original conception of the house, without curtains, would be realized." Adding insulation below the roof—there was none previously—and installing a new heating system helped make the old curtains unnecessary. Says Valentine, "We revealed the true lines of the house."

RENOVATION CHALLENGE "One mistake," says Matassoni, "was thinking we would want to spend even one night in the house listening to and smelling the old, mildewed air-blowing system." The couple didn't sleep in the house again until the new system was working.

RENOVATION DREAM Not only is this one of Philip Johnson's favorite houses, but it's only a few miles away from his own Glass House. That meant he was able to keep an eye on the renovation. On his last visit to the house, Johnson inscribed a book to Matassoni and Valentine: "I shall visit once a year to make sure it stays pure."

NATURAL SPIRIT

WHAT HE HAD A SMALL CABIN FILLED WITH CHILDHOOD MEMORIES (SOME GOOD, SOME BAD), ON THE EDGE OF WASHINGTON STATE'S YAKAMA INDIAN RESERVATION.
WHAT HE WANTED A HOUSE THAT REFLECTS HIS WORLDVIEW, IN WHICH EVEN THE HUMBLEST OBJECT IS TREASURED AND IN WHICH THE LINE BETWEEN INDOORS AND OUT IS JUST UNIMPORTANT.

Opposite: The wall-hanging in Leo Adams's living room—resembling a crucifix or a kimono—is the outside of an old refrigerator, which Adams found, rusted and shot at, in the desert. The dining table base is an old washtub resting on a wheel hub. Above: Walls are a mix of new wood and planks that Adams weathered on the outside of the house.

HOW HE GOT IT Like many renovations, Leo Adams's has never really ended. But in his case, it's not because the contractor stopped showing up, or because the tiles he wanted were lost in transit. Adams, a painter, sees his house as a canvas on which he is forever exploring techniques. A new phase of the renovation might begin when a friend sends him a present. Once it's unpacked, the wrapping could become a wall treatment or the surface of a table. A length of corrugated cardboard became a scalloped canopy over Adams's four-poster bed.

The house, on the border of the Yakama Indian reservation, in central Washington State, was once owned by Adams's grandfather, a tribal chief. But as a teenager, Adams wasn't welcome there. He says he was "highly creative, feminine, interested in furniture and flowers"—which made him anathema to his cattle-ranching clan. His refuge was the outdoors, where he found beauty in the subtle colors of the desert, and in nature's outcasts. A weed, says Adams, is just a flower that no one else has had the good sense to admire.

When his grandfather died, Leo Adams inherited the house and—in as much an act of exorcism as of design—began turning it into a home and painting studio. He literally broke down its four walls. Instead of pulling up to a front door, visitors arrive at a series of outdoor rooms, screened by trellises and weathered wood partitions and decorated with found objects. The transition from outdoors to indoors is gradual, as if the house needs time to assemble itself from the desert. "People say, 'The inside of your house is just like the outside'," reports Adams, obviously agreeing.

Eventually, an outside that feels enclosed gives way to an inside that feels open. Never boxed in, Adams even cut a hole in the kitchen floor, allowing him to borrow part of the house's crawl space for a conversation pit. Because the pit lowers furniture—and guests—beneath eye level, it makes the kitchen feel bigger than it is and creates a nook for after-dinner conversation. "All I needed was a saw," says Adams. That, and a surfeit of imagination.

Above (left): Adams hung an old washbasin over his kitchen's conversation pit, and used pieces of a rusted water heater for the fireplace surround. The stairway leads to two guest bedrooms. Opposite: Adams paints the "rugs" on sheets of Masonite and creates bouquets from weeds and sticks.

RENOVATION DREAM The kitchen is the heart of Adams's house, but that doesn't mean he needed expensive materials to make it special. For countertops, he used galvanized metal, which, he says, cost $300 total and, with its marblelike texture, "has more life in it than stainless." And he made the light fixtures out of butcher paper, which he treated with spray starch to bring out the paper's grain. (Inside, he used ordinary lightbulbs.) A chandelier is simply the same fixture with an added tier. ("It's a miracle we haven't burnt the place down," Adams admits.) He stores old pottery, including Bennington spatterware, in galvanized boxes. The wall behind them is stained plywood to which Adams attached two lengths of wire fencing, suggesting a Japanese screen—at negligible cost.

What the Pros Know About
Humble Materials

Adams covered the walls in gray army blankets instead of toile and upholstered Louis-something chairs in cotton duck instead of velvet. Area rugs are Masonite, painted by Adams to resemble marble or mosaic—but with its grain still showing through. (Adams's aesthetic demands that nothing pretend to be something it's not.) Seating includes banquettes covered in denim and felt. **Adams recycles materials from his own house.** When cedar siding on the exterior darkens with age, he brings it inside, where he uses it as paneling. He alternates the weathered boards with new plywood (which he "washes" with diluted latex paint). The result: interior walls with patterns that suggest expensive wainscoting or stonework. Furniture is left unfinished. The scalloped fringe on Adams's bed is corrugated cardboard.

RIGHTING WRIGHT

WHAT SHE HAD A FRANK LLOYD WRIGHT HOUSE WITH STUNNING GEOMETRY, BUT SHAG CARPET AND PAINT SO GARISH "YOU COULD SEE IT FROM THE AIRPORT 12 MILES AWAY."
WHAT SHE WANTED A GENTLE-HUED MASTERPIECE THAT WOULD BE AS MUCH A HOME AS A MUSEUM.

HOW SHE GOT IT Frank Lloyd Wright was America's greatest architect—but renovating one of his houses is no more glamorous than updating any other building. Linda Melton, the feisty Louisianan who bought this mountainside house in Phoenix, spent mornings on her hands and knees, washing squares of slate and handing them to the masons redoing her floors.

When Melton bought the place, it had more than its share of drawbacks: Floors were covered in shag carpet, woodwork had cigarette burns and the exterior walls (of concrete block) were painted an orange so bright, she says, "you could see it from the airport 12 miles away."

Still, the genius of Wright's architecture shone through. The house is at once ancient (it suggests American Indian kivas) and futuristic (is this the Jetsons' mountain hideaway?). In plan, the building consists of interlocking circles. The perimeter of the courtyard wall (the largest circle), intersects the center of the living room; the perimeter of the living room, the center of the kitchen; the perimeter of the kitchen, the center of the dramatic fireplace. Mil Bodron, a Dallas architect (and lifelong friend of Melton's), says the circles create a feeling of safety, but the way they interlock draws the eye from one point to the other. "There's harmony and energy at once," he says of the house, one of the last designed by Wright before he died in 1959.

At the start of the renovation, Melton worked with Taliesin Architects, the firm founded by Wright's disciples. But she wasn't interested in a faithful restoration. For the window frames, Melton wanted a coppery enamel, rather than the "Taliesin red" preferred by Wright. And she decided to install the russet-colored slate floor—in place of the concrete original—and cover the windows in custom copper screens. Her goal was to give the house a soft, feminine glow.

Eventually, she parted ways with Taliesin and brought in Bodron, who helped her furnish the house in a way that introduced Wright to other 20th-century masters: Wright's origami-like "Taliesin" chairs meet seating by Warren McArthur (their aluminum a perfect foil to the ubiquitous copper) and a curvy Angelo Donghia chaise. And, in keeping with Melton's desire for comfort, Bodron added banquettes covered in a plush, plum-colored fabric and piled with soft pillows. And why not? The more comfortable you are, the more likely you are to enjoy the architecture.

Previous spread: In Linda Melton's house, an interior "cornice" sweeps around the living room three feet below the ceiling. It conceals lighting, makes the perimeter seating areas feel intimate and gives the topmost windows the feeling of clerestories opening directly to the sky. Melton spent months looking for just the right finish for the house's metalwork before she settled on paint made for a '63 Chevy.

Clockwise (from top left): Frank Lloyd Wright created extraordinarily rich compositions from three elements—metal, mahogany and concrete block; in the foyer, Melton's collection of 1950s ceramics has pride of place on restored mahogany shelves; small guest rooms are like a ship's stateroom; the fireplace is a masterpiece of cantilevered curves, setting the tone for the entire house (it is flanked by an anthropomorphic tool set and a 1940s French torchère); in the master bedroom, architect Mil Bodron accentuated the feminine in Wright's curvy design with his furniture and fabric choices; coppery glass Bisazza tiles enliven Melton's shower (reflected in dressing room mirrors).

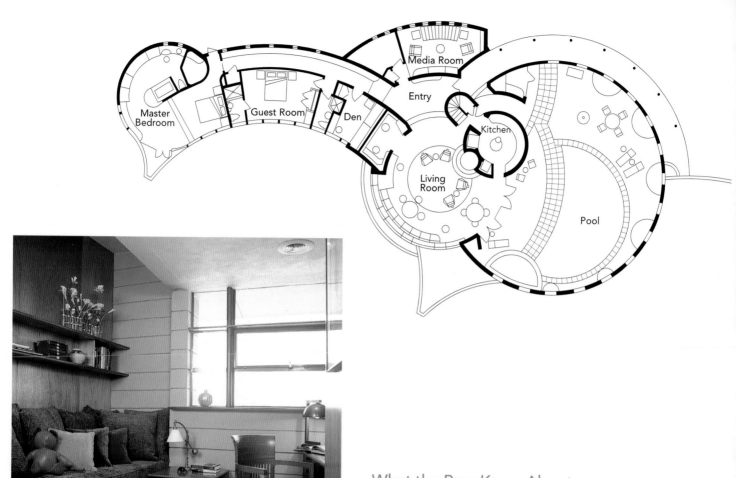

Master Bedroom · Guest Room · Den · Media Room · Entry · Kitchen · Living Room · Pool

What the Pros Know About
Window Screens

Window screens are energy savers—they let you open windows to cooling breezes when you might otherwise rely on air-conditioning. **But architects frequently avoid screens, which can complicate the task of creating sleek glass walls with a minimum of hardware. These days, screens are commonly made of dark gray fiberglass, which is inexpensive and nearly invisible.** But Linda Melton wanted to give her house a reddish glow, and she figured copper screens would help. TWP, a California company (twpinc.com), makes mesh in a variety of materials, including copper, stainless steel and bronze. Its products can be ordered in rolls, cut with a pair of scissors and inserted into window screen frames. Insect screens are generally 16 mesh (meaning 16 wires vertically and horizontally per inch). TWP sells 16 mesh copper for as little as $1.65 per square foot. The company's website advises that the copper is less durable than other screen materials and will inevitably weather. (Stainless steel, by contrast, is virtually indestructible and slightly less expensive.) But after several years, Melton says, her copper screens have developed "a gorgeous patina."

The half-moon windows in Melton's kitchen (opposite) overlook the pool; beyond the pool wall is a carport. Above the kitchen is a small library, its own windows "completing" the kitchen's. (Thanks to Melton, the half-moons, originally fixed, now open.) Bar stools aren't Wright—they're from New York's Dakota Jackson.

RENOVATION DREAM Melton first saw her house in a Frank Lloyd Wright book while she was raising a family in Louisiana. She says, "I dreamed that someday I would own it." Years later, on a golfing trip to Arizona, she drove by a sign that said "Frank Lloyd Wright house for sale." The realtor later told her that the sign had been up for less than an hour when she saw it.

RENOVATION CHALLENGE Melton wanted a pool, but the only place for one was inside the wall enclosing a small yard. But if the pool couldn't be big it could be beautiful. It was her idea to use handmade mother-of-pearl tiles that sparkle underwater. ("People kept telling me to use a pebbly brown surface," she says disdainfully.) For paving around the pool, Melton used more of the living room's slate, forging a strong connection between indoors and out (which was, after all, a hallmark of Wright's architecture). Architect Mil Bodron furnished the patio like a second living room. Half-moon cutouts in the paving contain plants that bring the desert into the architectural composition.

What the Pros Know About
Stainless-Steel Countertops

In much of her house, Linda Melton was going for a reddish glow. But for her kitchen counter, stainless steel was the Wright stuff. **Stainless can be cut, bent and welded into almost any shape—contractors make a substrate out of plywood, then a fabricator wraps the plywood in steel.** (A finished countertop will cost around $100 per square foot—custom edge details and integral backsplashes raise the price.) Since lengths of stainless steel can be welded together, and the weld marks ground down to invisibility, any size or shape is possible—just be sure there's a way to get it through a door or window. Mistakes can be fixed—Melton recalls that her counter arrived one day, didn't quite fit and came back the next day perfect. Stainless can last a lifetime. But oddly, despite its name, stainless has a maddening tendency to show fingerprints and even water "stains." As for scratches, they're inevitable. Most homeowners advise you not to worry about them—pretty soon, there will be so many of them that you can start referring to the ubiquitous marks as a "patina."

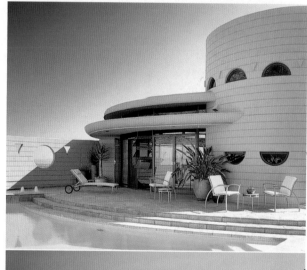

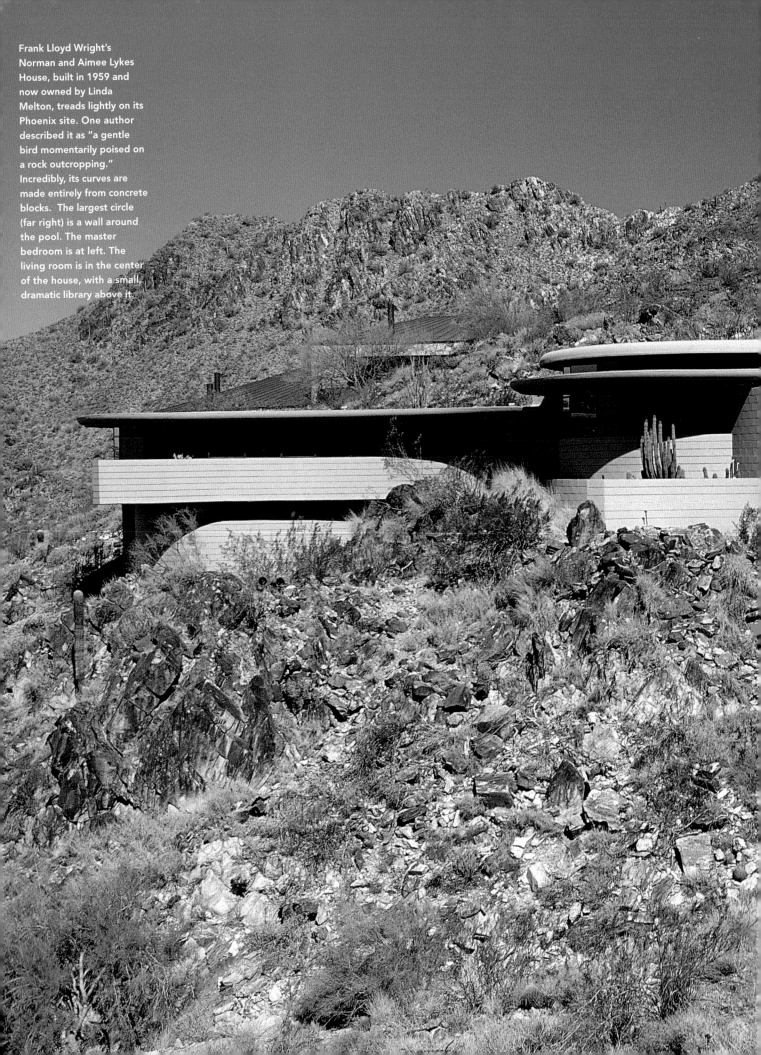

Frank Lloyd Wright's Norman and Aimee Lykes House, built in 1959 and now owned by Linda Melton, treads lightly on its Phoenix site. One author described it as "a gentle bird momentarily poised on a rock outcropping." Incredibly, its curves are made entirely from concrete blocks. The largest circle (far right) is a wall around the pool. The master bedroom is at left. The living room is in the center of the house, with a small, dramatic library above it.

RESOURCES

FLOOR SHOW

Design Nilus de Matran, Christopher Leitch & Dave Bailey, Nilus Designs, 757 A Pennsylvania Ave., San Francisco, CA 94107, 415/826-3434; **Contractor** John Hakewill Construction Co., 415/706-3873; **Engineer** Double-D Engineering, 415/551-5150; **Pages 10–11 Sofa** "Charles," by B&B Italia, 800/872-1697; **Lounge chairs** by Jasper Morrison, from Cappellini, 212/966-0669; **Stool** "Pebble," by Troy, 212/941-4777; **Fireplace surround, media cabinets** by George Slack Cabinetmakers, 415/285-0772; **Fireplace steel face** by Reification, 415/553-4183; **Windows** by Blomberg, 916/428-8060; **Flooring** by Associated Terrazzo, 415/641-1995; **Page 12 Coffee table** by Nilus Designs, by Paragon Frames, 415/552-7600; **Pendant lamps** "Karim Rashid's Soft Series," by George Kovacs Lighting, from City Lights, 415/863-2020; **Page 13 Sinks** by Just, from R.V. & Assoc., 707/745-3655; **Pot filler faucet** by Chicago Faucets, 847/803-5000; **Countertops** "Carrara Marble," from Luo Marble & Tile, 415/206-1819; **Cabinets** by George Slack Cabinetmakers, 415/285-0772; **Page 15 Dining table** custom by Elizabeth Paige-Smith, 310/455-3078; **Chairs** by Jasper Morrison, from Cappellini, 212/966-0669; **Page 16 Wallcovering, platform** by DecorAides, with Sophia Custom Drapery Workroom, 415/285-2344; **Floor covering** by Lonseal, from Ogden Contract Interiors, 415/824-9975; **Page 17 Studio table** by Nilus Designs, with Makerstudio, 510/452-3699; **Motorized shades** by Castec, from Sophia Custom Drapery Workroom, 415/285-2344; **Page 18 Bath tile** by Bisazza, from Ceramic Tile Design, 415/575-3785; **Vanity, cabinet** by George Slack Cabinetmakers, 415/285-0772; **Vanity top** from Lou Marble & Tile, 415/206-1819; **Vanity faucet** by Kroin at Hastings Tile, 212/674-9700; **Bathroom lighting** "Qua," by Foscarini, from City Lights, 415/863-2020; **Page 19 Bed** by DecorAides, with Sophia Custom Drapery Workroom, 415/285-2344; **Closets, storage** by George Slack Cabinetmakers, 415/285-0772.

UNCOMMON BRAHMIN

Architecture David Hacin, Hacin & Assoc., Inc., 46 Waltham St., Boston, MA 02118, 617/426-0077; **Design** Manuel De Santaren, MdS, Inc., 617/330-6998; **Page 20 Lounge chairs** "Nabab," from Holly Hunt, Chicago, 312/661-1900; **Page 21 Chair, far sofa, floor lamps** from Holly Hunt, Chicago, 312/661-1900; **End table** "Pigmee," and **Near sofa** "Nabab," from Holly Hunt, Chicago, 312/661-1900; **Cocktail table** from Carol Gratale, 212/838-8670; **Curtains** by Eliot Wright Workrooms, 617/542-3605; **Sofa table** by MdS, Inc., 617/330-6998; **Sconces** from Carol Gratale, 212/838-8670; **Pages 24–25 Cabinets** Painters' Collection, by SieMatic, 800/765-5266; **Page 26 Sink** "Memoirs," and **Tub** "Savoy," by Kohler, 800/4-KOHLER; **Page 27 Bed** by Antoine Citterio for Apta from Repertoire, 617/426-3865; **Bedside tables** by MdS, Inc., 617/330-6998; **Bedside lamps** by Stephane Davids from Repertoire, 617/462-3865; **Shades** from Eliot Wright Workrooms, 617/542-3605.

WORKING CLASSIC

Architecture Booth Hansen Assoc., Laurence Booth, Principal; Henry Soensken, Project Manager; 333 Desplaines, Chicago, IL 60661, 312/869-5000; **General contractor** Paul Zucker/City Real Estate, Inc., 773/525-3311; **Landscape** Maria Smithberg, Artemisia, 312/654-1708; **Page 28–29 Sofas, round side table** by B&B Italia, 800/872-1697; **Chair** "Costes," by Philippe Starck for Driade; **Ottomans** "Baleri," and **Coffee tables** all from Luminaire, 312/664-9582; **Fireplace** by Majestic, from Alltypes Fireplace + Stove Co., 708/383-6007; **Lighting** Lutron, fixtures by Juno, 800/323-5068; **Page 31 Table** "Mondo Moderno," by Piero Lissoni for Cappellini, at Luminaire, 312/664-9582; **Chairs** by Mario Bellini, from Luminaire, 312/664-9582; **Page 33 (bottom) Table** "Mondo Moderno," by Piero Lissoni for Cappellini, at Luminaire, 312/664-9582; **Page 34 Sink** by Kindred, 705/526-5427; **Faucet** by Grohe, 630/582-7711; **Page 35 Counters** by Chicago Marble Co., 630/734-1660; **Cabinet pulls** from Ironmonger, Inc., 312/527-4800; **Sink** by Kindred, 705/526-5427; **Faucet** by Grohe, 630/582-7711; **Stools** from Knoll, 212/343-4000; **Page 36 (top) Chair** from Knoll, 212/343-4000; **Bench** by Christian Liaigre, from Holly Hunt, 312/644-1728; **Sink, mirror cabinet** by Duravit, 888/DURAVIT; **Page**

36 (bottom) Bed by B&B Italia, from Luminaire, 312/664-9582; **Page 37 Pool** "Marathon Pool/Spa," from San Juan Pools, 800/535-7946; **Flooring** by Rossi USA Corp., 708/386-0183; **Wall tile** "Huang Hai," from Granite & Marble Resources, Inc., 312/670-4400; **Chair** from Knoll, 212/343-4000.

VISTA VICTORIAN

Architecture, design Anne Fougeron, Fougeron Architecture, 415/641-5744; **Page 38 Stools** Bertoia stools from Knoll, Inc., 800/445-5045; **Decorative steelwork** Dennis Leudeman, 510/658-9435; **Page 40 Leather chairs, coffee table** from the Mies van der Rohe Barcelona collection, **Green chair** Arne Jacobsen all at Knoll, Inc., 800/445-5045; **Sofa** Della Robbia, 949/251-8929; **Bookshelves** Fougeron Architecture, 415/641-5744; **Decorative steelwork** Dennis Leudeman, 510/658-9435; **Page 41 Table** Jasper Morrison, from Cappellini, 212/766-0669; **Chandelier** PH5 Pendant Lamp, from Louis Poulsen, 954/349-2525; **Page 42 Cabinets, Counters, Sink** Bulthaup, 800/808-2923; **Stools** Bertoia stools, from Knoll, Inc., 800/445-5045; **Pages 44–45 Cabinets** Fougeron Architecture, 415/641-5744; **Floor** Terrazzo by American Terrazzo, 415/921-1862; **Shower fixtures** "Tara," by Dornbracht, 800/774-1181.

CONDO MINIMAL

Architecture David Ling Architects, 225 E. 21st St., New York, NY 10010; 212/982-7089; **Window shades** by Hunter Douglas, 800/937-STYLE; **Page 46 Fireplace surround, backlit shield over fireplace** by David Ling Architects, 212/982-7089, fabrication Loye and Derikson, 845/255-4211; **Chair** "Lounge," by Eames, available at the MoMA Design Store, 800/793-3167; **Coffee table** by Gae Aulenti through Luminaire, 800/494-4358; **Armchairs** by Le Corbusier, through Cassina, 631/423-4560; **Page 48 Table** by Le Corbusier through Cassina, 631/423-4560; **Chairs** by Arne Jacobsen, through Fritz Hansen in Denmark, 45/48172300; **Cabinetry** by David Ling Architects, 212/982-7089; **Faucet** by Kroin at Hastings Tile, 212/674-9700; **Sink** by Elkay, 630/574-8484; **Page 49 Cabinetry** by David Ling Architects, 212/982-7089; **Page 50 Cabinetry, headboard** by David Ling Architects, 212/982-7089; **Ceiling hinge** by David Ling Architects, 212/982-7089; **Page 51 Tile** "Cobalt Blue Mosaic," by Hastings Tile, 212/674-9700; **Tub** by Kohler, 800/4-KOHLER; **Faucet** by Kroin at Hastings Tile, 212/674-9700.

BACK STORY

Architecture Gwynne Pugh, Pugh + Scarpa, 2525 Michigan Avenue, Santa Monica, CA 90404, 310/828-9532; **Design** Tim Clarke Design, Inc., 310/659-1950; **Contractor** Binder Minardo Builders, 310/828-6266; **Landscape architect** Lisa Mosely, 310/586-1178; **Pages 54–55 Sofas** by Tim Clarke Design, Inc., 323/656-1950; **Club chairs** from Modern One, 323/651-5082; **Side table, sofa, lamp** from Shelter, 323/937-3222; **Page 57 Chairs** "Saarinen," from Downtown, 310/652-7461; **Dining table** by Tim Clarke Design, Inc., 310/659-1950; **Page 58 Dining table** "Apta," from Diva, 310/278-3191; **Dining chairs** "Stacking chair," by Donghia, 310/657-6060; **Page 61 (top) Cocktail table** from Downtown, 310/652-7461; **Page 62 Bed** by Tim Clarke Design, Inc., 310/659-1950; **Love seat, chair** from Modernica, 323/933-0383; **Side lamp, bedside lamps, console** from Downtown, 310/652-7461; **Side table** from Blackman Cruz, 310/657-9228; **Page 63 Tub** by Kohler, 800/4-KOHLER; **Door manufacturer** by Fleetwood USA, fleetwoodusa.com.

LOST AND FOUND

Architecture Mell Lawrence Architects, 512/441-4669; **Interior design** Fern Santini, Abode, 512/300-2303; **Contractor** Kimberly Renner, The Castle Company, 512/469-5993; **Landscaping** James David, Gardens, 512/467-9934; **Construction consultant** Jerre Santini Construction, 512/751-2191; **Pages 64–65 Chandelier** "Flotation," by Ingo Maurer, through Abode, 512/300-2303; **Page 66 (left) Club chairs** Thomas O'Brien, for Hickory Chair, 828/328-1801; **Page 66 (right) Stools** the MoMA design store, 800/793-3167; **Page 67 Table** Rosewood Parsons table through Aqua 20th Century Modern, 512/916-8800; **Coffee table** Gardens, 512/451-5490; **Page 68 Bench** Gardens, 512/451-5490; **Page 69**

Bed by Rex D. White, Designer Craftsman, 830/997-2873; **Lamp** Ralph Lauren Home Collection, 212/642-8700; **Chair** "Lazy chair," by Zanzibar, 512/472-9234; **Metal door** The Castle Company, 512/469-5993; **Page 70 (left) Fixtures** by Red Dot, through Home Depot, 800/553-3199; **Page 70 (right) Indonesian chest** Gardens, 512/451-5490; **Sinks** Kohler, 800/4-KOHLER; **Page 71 Stool** Gardens, 512/451-5490; **Fixture** Kohler, 800/4-KOHLER; **Tub** The Castle Company, 512/469-5993.

GOOD AS GOLD
Landscape design John Byrd Garden Design, 704/377-0056; **Furniture** all furniture by Mitchell Gold, 800/789-5401, unless otherwise noted; **Page 72 Console, side tables** from Crate & Barrel, 800/323-5461; **Page 74 (bottom) Steel-glass table** from Storehouse, 404/233-4111; **Page 75 Side table** from Crate & Barrel, 800/323-5461; **Page 77 Counter, cabinets** by Grayfield Cabinet Co., 828/459-1144; **Island countertop** by Avonite, 505/864-3800; **Black countertops** by Wolf Appliance Co., 608/271-2233; **Island lighting** from Smith & Hawken, 800/776-3336; **Bar stools** from Storehouse, 404/233-4111; **Page 79 (bottom right) Bed, desk** by Baronet, 418/387-5431.

CAPITAL IMPROVEMENT
Interior design Mary Drysdale, Drysdale, Inc., 78 Kalorama Circle, NW, Washington, D.C., 20008, 202/588-0700; **Pages 80–81 Table** by Le Corbusier, from Cassina, 631/423-4560; **Painted chairs** "U.S. Government Issue," by Sam Gilliam, 202/588-0700; **Carts** from the MoMA design store, 800/793-3167; **Chair** "Painted Wooden Chair," by Gerrit Rietveld, from Cassina, 631-423-4560; **Page 82 Chair** "Le Bambole," from B&B Italia, 800/872-1697; **Tavern table** by Drysdale, Inc., 202/588-0700; **Floor painting** by the Billet Collins Studio, 301/670-5550; **Page 83 (top) Sofa** "Le Bambole," from B&B Italia, 800/872-1697; **Coffee tables** "Tree stump" tables, by Drysdale, Inc., 202/588-0700; **Page 83 (bottom) Andirons** "Sunflower," from Auffray and Co., 212/889-4646; **Finial** by Mary Drysdale and Sam Gilliam, 202/588-0700; **Page 84 Skylights** velux.com; **Page 87 Bed** by Drysdale, Inc., 202/588-0700.

WITHOUT BORDERS
Interior design Frankel + Coleman, 312/697-1620; **General contractor** Thorne Assoc., Inc., Chicago, IL; **Millwork contractor** Mielack/Woodwork, Edison, NJ; **Electrical contractor** Shamrock Electric Co., Inc., Chicago, IL; **Plumbing, heating contractor** Ashland Plumbing & Heating Co., Chicago, IL; **Pages 90–91 Sofas** Le Corbusier, from Cassina, 631/423-4560; **Coffee table** from Knoll, 212/343-4000; **Kyoto table** by Frattini, from Knoll, 212/343-4000; **Bar stools** "Zig Zag," by Gerrit Rietveld, from Cassina, 631/423-4560; **Track lighting** from Juno, 847/827-9880; **Page 94 (top) Dining table base** Le Corbusier, from Cassina, 631/423-4560; **Dining table tops** custom design by Neil Frankel, fabrication by Mielack/Woodwork; **Dining chairs** "Brno" chairs, by Mies van der Rohe; **Lounge chairs, chaise, table** by Mies van der Rohe, from Knoll, 800/343-5665; **Bed** custom design by Neil Frankel, fabrication by Mielack/Woodwork; **Page 94 (bottom left) Stools** Le Corbusier, from Cassina, 631/423-4560; **Kitchen table** from Zographos; **Child's table** by Alvar Aalto, from ICF, 800/237-1625; **Child's chair** "Model No. 69," by Alvar Aalto, from ICF, 800/237-1625; **Page 94 (bottom right) Bar stools** "Zig Zag," by Gerrit Reitveld, from Cassina, 631/423-4560; **Page 95 (left) Kyoto table** by Frattini, from Knoll, 212/343-4000; **Page 95 (right) Window shades** by Mechoshade, 718/729-2020, or through Marvin Feig & Associates, Chicago, IL; **Console table** by Mario Botta, through Luminaire, 800/494-4358; **Page 96 Fireplace** by Majestic, from Alltypes Fireplace & Stove Co., 708/383-6007; **Page 97 Fixtures** from Kohler, 800/4-KOHLER; **Lavatory top** design by Neil Frankel; **Cabinets** custom design by Neil Frankel, fabrication by Mielack/Woodwork.

DRAWING ROOM
Architecture Deborah Berke, Deborah Berke Architect, 211 W. 19th St., New York, NY 10011, 212/229-9211; **Design** by Thomas O'Brien, Aero Studios Ltd., 132 Spring St., New York, NY 10012, 212/966-4700;

Contractor J. Grace Co., 212/987-1900; **Pages 98–99 Sofas** from Jonas Upholstery, 212/691-2777; **Chairs** from Jonas Upholstery, 212/691-2777; **Fireplace tools** by John Boone, 212/758-0012; **Table, floor lamp** from Aero Ltd., 212/966-1500; **Stools** from Profiles, 212/689-6903; **Page 100 Sofas, chairs** from Jonas Upholstery, 212/691-2777; **Coffee table** "Maxalto," by B&B Italia, 212/758-4046; **Stools** from Profiles, 212/689-6903; **Page 101 Dining table** "Less," by Jean Nouvel, from Unifor, 212/673-3434; **Chairs** "Melandra," by B&B Italia, 212/758-4046; **Credenza** "Apta," from B&B Italia, 212/758-4046; **Page 102 Chair** "Harry," by B&B Italia, 212/758-4046; **Tables** by Aero Ltd., 212/966-1500; **Cubby storage, panel doors** by Deborah Berke Architect, 212/229-9211; **Page 103 Materials manufacturer** American Acrylic, americanacrylic.com; **Page 103 (top) Pocket doors** by Deborah Berke Architect, custom by Edelman Metalworks, 203/744-7331; **Page 103 (bottom) Tub, showerheads** by Waterworks, 212/371-9266; **Flooring** by Stone Source, 212/979-6400; **Pages 104–105 Sofa** "Harry," by B&B Italia, 212/758-4046; **Chairs, coffee table** from Aero Ltd., 212/966-1500; **Side table** "Thompson," Thomas O'Brien Collection for Hickory Chair, from Aero Ltd., 212/966-1500; **Storage-bin wall** by Pyramid Steel Shelving, 718/381-5770.

BOSTON BEACON
Architecture David Hacin, Hacin + Associates Architects, 617/426-0077; **Contractor** Paris Building Group, 781/932-9922; **Window film** CHB Industries, chbwindowfilm.com; **Page 106 Dining table** "Mondo," Repertoire, 617/426-3865; **Dining chairs** "Classica," Repertoire, 617/426-3865; **Sofa** "Swain," On The Fringe, 617/269-0859; **Ottoman** "Minotti," The Morson Collection, 617/482-2335; **Leather chair** "Moroso," Repertoire, 617/426-3865; **Coffee table** "Pigreco," Bmodern, 153 Wooster St., New York, NY 10012; **Page 107 (right) Lounge chairs** "Meridian chaise," Brown Jordan, 800/743-4252; **Page 109 Kitchen** SieMatic, 617/423-0515; **Page 112 (top) Bench** "Mood," Vibieffe 2000, Sedia, Inc., 617/451-2474; **Stool** "The Covey Stool," Machine Age, 617/482-0048; **Page 112 (bottom) Bed** "Escale," Adesso, 617/451-2212; **Page 113 Cabinetry, sink, tub surround** Kochman, Wreidt and Haigh Cabinet Makers, 781/341-4313; **Faucet** Dornbracht, 800/774-1181; **Light fixture** "Flou," Chimera Lighting, 888/444-1812; **Chair** Portico, 212/941-7800.

LIGHT INDUSTRY
Architecture Jennifer Luce, Principal, and Sharon Stampfer, Associate, Luce et Studio, 1037 J St., San Diego, CA 92101, 619/544-0223; **General contractor** Mark Stangl Construction, 619/669-0800; **Structural engineer** Ishler Engineering, 310/581-8486; **Electrical** Zed Electric, 619/224-2748; **Metal fabricators** Alpine Metal Works, 619/445-8802; Benchmark Custom Welding, 619/523-1387; GoppWerks, 858/566-8710; **Custom cable** Rigworks, 619/223-3788; **Cabinetry** Jacobs Woodworks, 619/293-3702; John Pierson, 912/898-0382; Gary Gilbert, 760/720-0965.

MYSTERY HISTORY
Architecture, design Kevin Walz, Walzwork, Inc., 315 W. 36th St., Suite 1000, New York, NY 10018, 212/736-2399; **Contractor** Gerry Gallagher, Glebe Construction, 914/234-2781; **Landscape, design** Charles Marder, 631/537-3700; **Pages 124–125 Daybed** from Paula Rubenstein Antiques, 212/966-8954; **Ottoman, sofa, side chair, armless chair** from Pembroke & Ives, 212/995-0555; **Tea table, coffee table, foot stool, silversmith table, mahogany chairs** from Anglo Raj Antiques, 212/689-3437; **Side lamp** from Ann Morris Antiques, 212/755-3308; **Page 126 Hanging lamp, chair** from Ann Morris Antiques, 212/755-3308; **Page 127 (right) Table, chairs** from Anglo Raj Antiques, 212/689-3437; **Hanging lamps** from Ann Morris Antiques, 212/755-3308; **Page 128 Custom lighting** by Fran Taubman, 631/537-3579; **Lamp** from Anglo Raj Antiques, 212/689-3437; **Coffee table** from American Wing, 631/537-3319; **Page 130 (left) Cabinet, lavatory cabinet, wall shelf** by Kevin Walz, Walzworks, Inc., 212/736-2399; **Painting** by Amy Kaiser-Wickersham, 914/381-2718; **Sconces** from Ann Morris Antiques, 212/755-3308; **Bench** from Anglo Raj Antiques, 212/689-3437; **Page 131 Bed, table** from Anglo Raj Antiques, 212/689-3437; **Lamps** by Kevin Walz, from Baldinger Lighting, 718/204-5700.

MIAMI TWICE

Architecture Warren Ser, Ser Design Assoc., Inc., 305/762-7533; **Interior architecture** Alison Spear, Alison Spear, AIA, 305/438-1200; **Project manager** Thaïs Fontenelle, Alison Spear, AIA, 305/438-1200; **Contractor** Cliff Blate, Blate Construction Co., Inc., 305/665-8760; **Landscape architecture** Robert Parsley III, Geomantic Design, 305/665-9688; **Lighting design** Sylvia Bastrong, ISP Design Team, 305/278-1565; **Page 132 Sofa** "Le Canapé," by Flexform, at Luminaire, 800/494-4358; **Candelabrum** Christian Astugueveille for Holly Hunt; 305/571-2012; **Page 134 (left) Fireplace mantel** Alison Spear, AIA, 305/438-1200; **Page 134 (right) Table** Knoll, 800/445-5045; **Chairs** Luminaire, 800/434-4532; **Hanging fixture** Senzatempo, 305/534-5588; **Page 137 Countertops** Coverings, 305/572-1080; **Table** Monica Armani Progetto 1, at Luminaire, 800/434-4532; **Chairs** Panama chairs by Antonio Citterio, at Luminaire, 800/434-4532; **Page 138 Sinks** Kohler, 800/4-KOHLER; **Marble** Coverings, 305/572-1080; **Fixtures** Dornbracht, 800/774-1181; **Tub** Agape "Spoon," by Carlo Benedini, at Luminaire, 800/434-4532; **Waterproofing membrane system** by Laticrete, laticrete.com; **Page 139 (middle row, left) Wicker chairs** "Apollo" chairs, by Ross Lovegrove, at Luminaire, 800/434-4352; **Aluminum chairs** Martin Van Severen, at Luminaire, 800/434-4532; **Chaise** Schultz 1966, at De Greco, 121/688-5310; **Page 139 (middle row, center) Bed** Alison Spear, AIA, 305/438-1200; **Side tables** Cappellini Sistemi, at Luminaire, 800/434-4532; **Lamp** Chris Harty, 305/674-8629 **Page 139 (bottom row, left) Side tables** Gilbert Rhode for Chris Harty, 305/989-7431; **Cabinets** Alison Spear, AIA, 305/438-1200.

PICTURE WINDOW

Architecture David Lake, Lake/Flato Architects, 311 Third St., San Antonio, TX 78205, 210/227-3335; **Landscaping** Rosa Finsley, King's Creek Nursery, 214/878-3975; **Hardscape** Big Red Sun, 1627 Willow, Austin, TX 78702, 512/480-9749; **Windows** marvin.com; **Page 140 Armless chairs** from A.K. 11 14, 310/399-1453; **Sofa** from the owners' collection, rebuilt by Rudy's Custom Upholstery and Design, 210/821-5156; **Birdbath** by David Lake, Graham Martin, Jay Hargrave, Lake/Flato Architects, 210/227-3335; **Page 141 Dining table** by Billy Johnson, Lake/Flato Architects, 210/227-3335; **Chairs** by Cassina, at Ferguson-Rice, 713/666-8585; **Sofa, hassock** from the owners' collection, rebuilt by Rudy's Custom Upholstery and Design, 210/821-5156; **Side table** from Jean-Marc Frey French Antiques and Decorative Art, 512/474-4660; **Page 144 Bar stools** "Pila," by ICF, 800/237-1625, through Sue Gorman, 713/621-1988; **Lighting** "Island Light Fixtures," by Graham Martin and David Lake, Graham Martin Lights, 210/490-9654; **Page 147 Front door** by David Lake and Alex de Leon, built by Alex de Leon, 210/224-1160.

BARN AND NOBLE

Architecture, interior design D'Aquino Monaco, Inc., 180 Varick Street, New York, NY 10014, daquinomonaco.com, 212/929-9787; **Page 150 Coffee table** Burning Relic, 718/625-5880; **Art deco chaise, art deco chairs** Two Zero Applied Art, London, 011+44+207/720-2021; **Page 153 Circular breakfast tables**, Tucker Robbins, 212/366-4427; **Leather breakfast chairs** Cassina, USA, 631-423-4560; **Cabinetry** Custom designed by D'Aquino Monaco, Inc., 212/929-9787; **Page 155 Four-poster bed**, Ralph Lauren Home Collection, 212/421-1200; **Cabinetry** Custom designed by D'Aquino Monaco, Inc., 212/929-9787.

SCENE STEELER

Architecture Tom Kundig, FAIA, Olson Sundberg Kundig Allen Architects, 159 South Jackson Street, 6th Floor, Seattle, WA 98104, 206/624-5670; **Contractor** David Boone, Boone Construction, 206/378-1169; **Floors** Environmental Home Center, 800/281-9785; **Page 156 Bench** "Gallery seat," from Palazetti, 888/881-1199; **Coffee tables** from IKEA, 800/434-IKEA; **End table** from Pottery Barn, 800/922-5507; **Lighting** by Max Moseley, 206/313-6759, Andrew Elliott, 206/295-8908; **Page 158 Console** "Xenox," from Limn, 415/543-5466; **Page 159 Dining, table, benches** from David Smith & Co., 206/223-1598; **Counters** by Dog Paw Design, 206/706-0099; **Stools** from Limn, 415/543-5466; **Lighting** by Elliott Bay

Electric, 206/660-4006; **Locker cabinets** by Lyon from Engineered Storage Products, 206/682-6596; **File cabinets** from Sonrisa, 323/935-8438; **Glass shelves** by Tom Kundig, Jim Graham, built by Boone Construction; **Faucets** by Chicago Faucets, 847/803-5000; **Page 160 Bed** "Claudiano," from Limn; **Page 161 Sinks** by Villeroy & Boch, 212/677-1151; **Backsplash** by Tom Kundig, Jim Graham; **Sconces** by Current, 206/622-2433; **Lighting** by Elliott Bay Electric, 206/660-4006; **Tub** "New Haven," by Villeroy & Boch, 212/677-1151.

HIGH BEAMS

Interior design Tori Golub, Tori Golub Interior Design, 26 E. 63rd Street, New York, NY 10021, 212/583-9570; **Renovation consultants** Hanes-Wohl Designs, 310/289-9559; **Art consultants** Winston-Wachter Fine Arts, 212/327-2526; **Page 164 Club chairs, ottomans** from City Antiques, 323/658-1085; **Nesting tables, African stool** from Dragonette Decorative Arts, 310/855-9091; **Fireplace mantel** by Tori Golub Interior Design, 212/583-9570; **Mantel fabrication** by Eddie Carillo, 310/397-1588; **Page 165 Steel/glass table** from See, 212/228-3600; **Chairs** vintage from Jules Seltzer Assoc., 310/274-7243; **Suspended light** vintage, from A.K. Eleven Fourteen, 310/399-1453; **Daybed** by Chris Lehrecke, from Pucci International, 212/633-0452; **Cork coffee table** vintage from Russell Simpson Co., 323/651-3992; **Side table** by Eileen Gray, from Modern Living, 310/657-8775; **Page 167 Steel-glass table** from See, 212/228-3600; **Chairs** from Jules Seltzer Assoc., 310/274-7243; **Suspended light** from A.K. 11 14, 310/399-1453; **Buffet** from Wyeth, 212/925-5278.

HIGH STYLE

Design John Saladino, Saladino Furniture, Inc., 212/684-3720; **Head designer for the project** Alan Sendelbach, 212/684-6805; **Pages 170–171 Sofas, coffee table, ottomans, side chair, screen** by John Saladino, Saladino Furniture, Inc., 212/684-3720; **Page 174 Sofa, armchair** by John Saladino, see above; **Wall behind sofa** SJW Studios Inc. through Kirk Brummel Wallpaper, 212/477-8590.

MUSEUM QUALITY

Architecture, design Julie Dowling, Dowling Kimm Studios, 415/241-0700; **Contractor** Dean Patyk Construction, 415/254-3834; **Walnut floors** Zec Hardwoods, 415/924-5456; **Stone** Sterner Marble & Granite, Inc., 510/215-1866; **Tile** Moppert Tile, 415/998-4909; **Cabinetry** Apple Woodworks, 707/525-1978, Ken Rose, 415/669-7660; **Stainless-steel railings** Nueva Castilla, 415/282-6767; **Lighting/electrical** Marina Electric, 415/885-3604; **Integrated Audio System** Performance Audio, 415/441-6220; **Pages 176–177 Sofa, chairs, chaise** from Lounge, 415/563-2220; **Lamps** Modenature, from Bee Market, 415/292-2910; **Coffee table** Catherine Memmi, from Bee Market, 415/292-2910; **African daybed** Antique from Ron Messick Fine Art, 505/983-9533; **Page 179 (top left) Table** from Christophe Delcourt (Paris Showroom), 011+33 1/42-78-44-97; **Page 179 (center left) Sofa, chairs, ottoman** from Lounge, 415/563-2220; **Side tables** by Eileen Gray from Design Within Reach, 800/944-2233; **Page 179 (center right) Faucets** Dornbracht from DJ Mehler, 415/490-5742; **Vanity, cabinets** Inda from Arkitektura, 415/565-7200; **Page 179 (bottom) Silk pendant fixture** custom by Bone Simple Design, 212/627-0876; **Table** Home from Bee Market, 415/292-2910; **Cabinetry** design by Julie Dowling, Dowling Kimm Studios, 415/241-0700, fabrication by Apple Woodworks, 707/525-1978; **Limestone countertops** Sterner Marble & Granite, Inc., 510/215-1866; **Ultrasuede banquette** design by Julie Dowling, Dowling Kimm Studios, 415/241-0700, from Lounge, 415/563-2220; **Stainless-steel cabinet pulls** from Bauerware, 415/864-3886; **Page 180 Stainless-steel railing** design by Julie Dowling, Dowling Kimm Studios, 415/241-0700, fabrication by Nueva Castilla, 415/282-6767; **Ebonized sculpture niches** design by Julie Dowling, Dowling Kimm Studios, 415/241-0700, fabrication by Ken Rose, 415/669-7760; **Pages 182–183 Vertical bamboo cabinetry** design by Julie Dowling, Dowling Kimm Studios, 415/241-0700, fabrication by Ken Rose, 415/669-7760, material by Ply-Boo, plyboo-plywood.com; **Ultrasuede bed** from Lounge, 415/563-2220; **Stool** Modenature from Bee Market, 415/292-2910.

SOFT TOUCH

Architecture Tsao & McKown Architects, 20 Vandam Street, NY, New York 10013, 212/337-3800, tsao-mckown.com; **Page 184 Rope chair** by Droog; **Side chair** by Edward Wormley; **Page 185 Sofa, coffee table, wall sconces, pocket door, dining table** by Tsao & McKown, 212/337-3800; **Ceiling fan** "Duplo Dynamico," from Lighting by Gregory, 212/226-1276; **Dining end chairs** "Dunbar," 1960, by Wyeth; **Page 187 (bottom right) Tile** from Bisazza, 305/597-4099; **Sink, fixtures, showerhead** from Waterworks, 800/998-BATH; **Cabinets, shower door** by Tsao & McKown, 212/337-3800; **Pages 188–189 Cabinets, bench, table, desk, chairs** by Tsao & McKown, 212/337-3800; **Pages 190–191 Lucite chair** by Shiro Kuramata; **Storage cabinets, bedside tables** by Tsao & McKown, 212/337-3800; **Reading lamp** from Ann Morris Antiques, 212/755-3308.

LOFT HORIZONS

Architecture Charles H. Bohl, Bohl Architects, 410/263-2200; **Curtain fabrication** by Mary Bright, 212/677-1970; **Page 194 Fireplace** KR-3 by Superior, 800/854-0257; **Vintage daybed upholstery** "Patina," to the trade by Pollack & Associates, 212/421-8755; **Curtain tracks** from Silent Gliss, 770/466-4811; **Lucite side table** Martha Sturdy, 604/872-5205; **Chair** Marcel Wanders's "Knotted" chair at Cappellini; **Stool** Sori Yanagi "Butterfly" stool at MoMA design store, 800/793-3167; **Page 195 Bureau** Chris Lehrecke at Pucci, 212/633-0452; **Page 196 Table, chair** Chris Lehrecke at Pucci, 212/633-0452; **Curtain** to the trade "Icarus" fabric by Bergamo, 212/888-3333; **Page 197 Brackets** "Ellen's Brackets" by Parallel Design at Doug Mockett & Co., 800/523-1269; **Stool** Chris Lehrecke at Pucci, 212/633-0452; **Baseboard** Formica DecoMetal laminate, 800/367-6422; **Page 198 Kitchen Countertops** Formica Surell solid surfacing, 800/367-6422; **Cabinets** by Knosos, 212/242-0966; **Page 200 (left) Table** Vintage Nakashima through Lawrence Converso, 312/755-1761; **Chairs** Koi chairs at Troy, 212/041-4777; **Vintage chairs** Donzella, 212/965-8919; **Page 200 (right) Lamps** David Weeks at Pucci, 212/633-0452; **Page 201 Tiles** Vidrotil at Hastings, 516/379-3500, installed by Guri, 917/217-5925; **Glass door** Alumex, 800/521-2284; **Fixtures** by Kohler, 800/4-KOHLER.

SMALL MIRACLE

Architecture, design Steven Shortridge, Callas, Shortridge, 3621 Hayden Ave., Culver City, CA 90232; 310/280-0404; callas-shortridge.com; **Contractor** Michael Conn, 310/395-0018; **Painting contractor** Carl Tillmanns, 323/965-8518; **Landscape architect** Jay Griffith, 310/392-5558; **Page 202 Sofa** by Vladimir Kagan, from the Jordan Collection, 310/659-0771; **Small table** "Tulip," by Saarinen, from Jules Seltzer Assoc., 310/2747243; **Brown chair, Ottoman** "Grasshopper," by Saarinen, at Modernica, 323/933-0383; **Page 203 Exterior plaster** Steel-troweled, color integral plaster, by Eddie Carillo, 310/397-1588; **Page 204 (top) Drafting table** by Ron Rezek; **Chair** by Eames, at Herman Miller for the Home, 800/646-4400; **Bookcase** "Expedite," from IKEA, 800/434-IKEA; **Page 204 (bottom left) Dining table** "Tulip," by Saarinen, from Knoll, 310/289-5800; **Dining chairs** by Eames, at Herman Miller for the Home, 800/646-4400; **Page 204 (bottom right) Built-in cabinet** by Steven Shortridge, 310/652-8087, fabricated by Spiral III Design, 323/733-9951.

NEW LEASE

Architecture, design Martha Finney, 215/483-5241; **Siding** American Cedar Millwork, 302/645-9580; **Windows** Grubb Lumber Co., 302/652-2800; **Concrete products** Kemiko, kemiko.com; **Pages 208–209 Tabletop** Fireslate, 800/523-5902; **Chairs** Sam Timberlake, 207/824-2234; **Cabinets** Fiorella Woodworking, 215/843-5870; **Page 212 Shower curtain rod** General Cubicle Co., Inc., 215/723-8932; **Page 213 (right) Headboard** Garrett Finney, FARO, 713/303-3862.

PERFECT FIT

Architecture, design Principal in charge, James Gauer; Project architect, Mariana Marron, Gauer & Marron Studio, 212/417-9198; **Contractor** Miro Balac, Miro Balac, Inc., 914/771-5830; **Cabinetry** Wolfgang Micheltsch, Euro Woodworking, 917/861-5369; **Material** Fin Color Ply, fincolorply.com/clrply; **Steel door and window fabrication** Alan Herman, A & S Steel Windows, 718/275-7900; **Stainless-steel fabricator** Peter Chang, Bowery Restaurant Supply, 212/228-2900; **Painter** Dragan Urosevic, Noble Housing Painting, 718/457-1756; **Tile installer** Branko Zec, Zec Enterprises, Inc., 917/282-0416; **Electrical contractor** Warren Ostroff, Ostroff Electric, Inc., 718/323-5200; **Heating and AC contractor** Tommy Nakahashi, Nakahashi Mechanical Contracting, Inc., 201/227-9799; **Pages 216–217 Sofas** custom through Dune, 212/925-6171; **Coffee tables** "Hoffmann Nesting Tables #100 T00 EB," at ICF, 212/388-1000; **Sconces** "Christian Liaigre Arthur Sconce #ARTO-SCH-1," through Holly Hunt, 800/229-8559; **Credenza** custom by Gauer & Marron Studio, 212/417-9198; fabricated by Euro Woodworking, 917/861-5369; **Table, chairs on deck** "Pinot Bistro Table #AR0172" and "Aero Chair #AR0602" both through Design Within Reach, 800/944-2233; **Pages 218–219 Chairs** "Madison chair" by Desiron, 212/353-2600; **Window shade** Janovic Plaza, 212/595-2500; **Page 220 Cabinetry** by Peter Chang, Bowery Restaurant Supply, 212/228-2900; **Sliding door** custom by Gauer & Marron Studio, 212/417-9198; fabricated by A & S Steel Windows, 718/275-7900; **Page 222 (top) Tiles** "Fifth Avenue Collection," by United States Ceramic Tile, 330/866-5531; **Sink** "Happy D Basin/Console," by Duravit; **Fixtures** "Tara," by Dornbracht; **Shelf brackets** "KRT25HLD-016," by Kroin all through Quintessentials, 212/877-1919; **Light fixture** "SC71," by Modulighter, Inc., 212/371-0336; **Page 223 Shelves/cabinetry** custom by Gauer & Marron Studio, 212/417-9198; fabricated by Euro Woodworking, 917/861-5369; **Lamp** by GL Lites On, 212/534-6363, all recess lighting by GL Lites On.

HONORING PHILIP JOHNSON

Glass contractor Joe Schiano, House Medic, 203/829-1760; **Glass maintenance** Ken Ray, 203/301-0699; **Roofing** Bill Eleck, Eleck & Co., 203/831-0691; **Heating/cooling** Kevin Carrey, Petro, Inc., 800-OIL-HEAT; **Landscape lighting** Mark Nielson, Nielson Electric, 203/869-8184; **Pages 226–229 Furniture** all from Knoll, 212/343-4000; **Page 230 Dining table** from Pace, 212/838-0331; **Page 231 Chairs** from Knoll, 212/343-4000.

NATURAL SPIRIT

Design Leo Adams, 509/966-5900. Leo Adams's unique house is furnished with antique found objects and one-of-a-kind pieces.

RIGHTING WRIGHT

Interior design Mil Bodron Design, Bodron & Fruit, Inc., 4040 North Central Expressway Suite 150, Dallas, TX 75204, 214/826-5200, bodronfruit.com; **Contractor** Ron Steege, Tim Larson, La Casa Builders, 480/404-0105; **Landscape architecture** Jeff Berghoff, Marcus Bollinger, Berghoff Design Group, 480/481-3433; **Pages 242–243 Lounge chairs** "Taliesin"; **Dining chairs** "Barrel," both by Frank Lloyd Wright, at Cassina, 631/423-4560; **Armchairs** at Fred Silverman, 212/925-9470; **Stools** by Frank Lloyd Wright; **Bronze table** by Gene Summers, at Holly Hunt, 312/644-1782; **Page 246 (top right) Ceramics** 1950s Swedish, through Lin/Weinberg Gallery, 212/219-3022; **Page 246 (bottom right) Bed** by Berman/Rosetti, at David Sutherland, 214/742-6501; **Page 247 (top) Stools** by Frank Lloyd Wright; **Chairs** "Barrel," both by Frank Lloyd Wright, at Cassina, 631/423-4560; **Lamp** from William Lipton, 212/751-8131; **Page 249 (bottom) Chaise, chairs, tables** by Helzer, at David Sutherland, 214/742-6501; **Page 249 (top) Bar stools** "Saturn Stool," by Dakota Jackson, 212/838-9444, at David Sutherland, 214/742-6501.

Items that are not credited are antiques, vintage or from the owner's collection.

Because the rooms featured on these pages are in private homes, all the items pictured may not correspond to those available in our Resources guide. We offer only an approximate listing of widely accessible shops where the likelihood of finding similar items is greatest. The author and the publisher do not endorse any of the products mentioned or pictured.

CREDITS

Cover: Produced by Linda O'Keeffe; Photograph by Antoine Bootz.
Floor Show Photographs by Tim-Street Porter; **Uncommon Brahmin** Produced by Doretta Sperduto & Donna Paul; Photographs by Jeff McNamara; **Working Classic** Produced by Linda O'Keeffe & Lisa Skolnik; Photographs by Antoine Bootz; **Vista Victorian** Produced by Diane Dorrans Saeks; Photographs by Grey Crawford; **Condo Minimal** Produced by Linda O'Keeffe; Photographs by Maura McEvoy; **Back Story** Produced by Linda O'Keeffe & Laura Hull; Photographs by Jeremy Samuelson; **Lost and Found** Produced by Doretta Sperduto & Helen Thompson; Photographs by Jeff McNamara; **Good as Gold** Produced by Doretta Sperduto; Photographs by Joshua McHugh; **Capital Improvement** Produced by Doretta Sperduto & Barbara Bohl; Photographs by Celia Pearson; **Without Borders** Produced by Linda O'Keeffe & Lisa Skolnik; Photographs by Antoine Bootz; **Drawing Room** Produced by Linda O'Keeffe; Photographs by Laura Resen; **Boston Beacon** Produced by Doretta Sperduto & Donna Paul; Photographs by Peter Murdoch; **Light Industry** Produced by Linda O'Keeffe; Photographs by Jeremy Samuelson; **Mystery History** Produced by Doretta Sperduto; Photographs by Pieter Estersohn; **Miami Twice** Produced by Linda O'Keeffe & Nisi Berryman; Photographs by Quentin Bacon; **Picture Window** Produced by Doretta Sperduto & Helen Thompson; Photographs by Maura McEvoy; **Barn and Noble** Produced by Linda O'Keeffe; Photographs by Catherine Tighe; **Scene Steeler** Produced by Doretta Sperduto & Linda Hunphrey; Photographs by Michael Jensen; **High Beams** Produced by Linda O'Keeffe; Photographs by Grey Crawford; **High Style** Produced by Linda O'Keeffe; Photographs by Antoine Bootz; **Museum Quality** Produced by Linda O'Keeffe; Photographs by Grey Crawford; **Soft Touch** Produced by Linda O'Keeffe; Photographs by Catherine Tighe; **Loft Horizons** Photographs by Antoine Bootz; **Small Miracle** Produced by Linda O'Keeffe & Laura Hull; Photographs by David Phelps; **New Lease** Produced by Linda O'Keeffe; Photographs by Catherine Tighe; **Perfect Fit** Produced by Linda O'Keeffe; Photographs by Catherine Tighe; **Honoring Philip Johnson** Photographs by Michael Luppino; **Natural Spirit** Produced by Doretta Sperduto & Linda Humphrey; Photographs by Jonn Coolidge; **Righting Wright** Photographs by Tim-Street Porter.

ACKNOWLEDGMENTS

Working with the extraordinarily talented editors of *Metropolitan Home*, including Donna Warner, Michael Lassell, Linda O'Keeffe and Arlene Hirst, is one of the great joys of my life. They have made discovering—and explaining, in words and pictures—how architects, designers and homeowners create meaningful living spaces a labor of love.

Donna Warner was the champion of this book, and her contributions at every stage cannot be overstated. She was joined by Dorothée Walliser, for whose tireless stewardship of the project I am extremely grateful. Keith D'Mello, art director of the magazine and of this book, was a remarkably helpful and supportive colleague. Jenine Iannello, Dan Golden, Andy Pearce, Lynn Messina, Emily Furlani, Annette Farrell, Kate Walsh, Susan Alter and others worked tirelessly to bring this project to fruition.

Each of the locations in this book appeared in *Metropolitan Home* in different form; I am grateful to the writers of those articles: Michael Lassell, Raul Barreneche, Joseph Giovannini, Jeff Book, Helen Thompson, Sarah Lynch, Arlene Hirst, Diane Dorrans Saeks, Jorge S. Arango, Sharon Donovan, Frank De Caro, Anna Kasabian, Beth Dunlop, Mindy Pantiel, Susan Kleinman and Lucie Young. I am honored to be part of this group.

I am also grateful to the photographers and the producers, including Linda O'Keeffe and *Metropolitan Home*'s extraordinary "city editors," who helped create the photos in this book.

The homeowners, architects and designers whose work is represented here were extraordinarily generous with their time; without them the book could not have happened.

Charles Upchurch encouraged me to find time for this project, and for his support (and the smiles of Aaron and Jacob) I am grateful beyond measure.

Fred Bernstein
New York City

This book could not exist without the amazing staff of *Metropolitan Home* magazine. The energy, creativity and caring of this tiny group (there are 11 of us in editorial and art) make *Met Home* an extraordinary place to work. I believe both the magazine and this book reflect our enthusiasm for our work and the professionalism we bring to our jobs. Special thanks, of course, go to Fred Bernstein, the magazine's senior contributing editor and a former member of the staff, for his masterly work as author of this book and to Keith D'Mello, art director extraordinaire.

Contributing and city editors and too many loyal writers and photographers to name are our far-flung network of eyes and ears. Without them, we'd never find so many wonderful homes. And we stand in awe of all "our" farsighted homeowners, designers and architects.

Many others are responsible for our pages. In our dozen years at Hachette Filipacchi Media U.S., two editorial directors have been patient mentors and enthusiastic cheerleaders, so many thanks to Jean-Louis Ginibre and François Vincens. Thanks also go to Jack Kliger, our CEO, for his staunch support of our modern vision, and of course to our publisher, Anne Triece, her staff, as well as all the people on our business side and to Dorothée Walliser of Filipacchi Publishing, who kept us on track with a superb sense of direction, tight focus and a great sense of humor.

My parents taught me that home is the most important place. My father, a restaurateur, never tired of making both his homes and his restaurants beautiful retreats. My husband, Niall, and my daughters, Jennifer and Vanessa (now with their own homes and families), and all our four-footed companions have always made home my favorite place to be. I hope this book (and those you love) make you feel the same.

Donna Warner
Editor in Chief, *Metropolitan Home*